Lecture Notes in Biomathematics

Vol. 1: P. Waltman, Deterministic Threshold Models in the Theory of Epidemics. V, 101 pages. 1974.

Vol. 2: Mathematical Problems in Biology, Victoria Conference 1973. Edited by P. van den Driessche. VI, 280 pages. 1974.

Vol. 3: D. Ludwig, Stochastic Population Theories. VI, 108 pages. 1974.

Vol. 4: Physics and Mathematics of the Nervous System. Edited by M. Conrad, W. Güttinger, and M. Dal Cin. XI, 584 pages. 1974.

Vol. 5: Mathematical Analysis of Decision Problems in Ecology. Proceedings 1973. Edited by A. Charnes and W. R. Lynn. VIII, 421 pages. 1975.

Vol. 6: H. T. Banks, Modeling and Control in the Biomedical Sciences. V, 114 pages. 1975.

Vol. 7: M. C. Mackey, Ion Transport through Biological Membranes, An Integrated Theoretical Approach. IX, 240 pages. 1975.

Vol. 8: C. DeLisi, Antigen Antibody Interactions. IV, 142 pages. 1976.

Vol. 9: N. Dubin, A Stochastic Model for Immunological Feedback in Carcinogenesis: Analysis and Approximations. XIII, 163 pages. 1976.

Vol. 10: J. J. Tyson, The Belousov-Zhabotinskii Reaktion. IX, 128 pages. 1976.

Vol. 11: Mathematical Models in Medicine. Workshop 1976. Edited by J. Berger, W. Bühler, R. Repges, and P. Tautu. XII, 281 pages. 1976.

Vol. 12: A. V. Holden, Models of the Stochastic Activity of Neurones. VII, 368 pages. 1976.

Vol. 13: Mathematical Models in Biological Discovery. Edited by D. L. Solomon and C. Walter. VI, 240 pages. 1977.

Vol. 14: L. M. Ricciardi, Diffusion Processes and Related Topics in Biology. VI, 200 pages. 1977.

Vol. 15: Th. Nagylaki, Selection in One- and Two-Locus Systems. VIII, 208 pages. 1977.

Vol. 16: G. Sampath, S. K. Srinivasan, Stochastic Models for Spike Trains of Single Neurons. VIII, 188 pages. 1977.

Vol. 17: T. Maruyama, Stochastic Problems in Population Genetics. VIII, 245 pages. 1977.

Vol. 18: Mathematics and the Life Sciences. Proceedings 1975. Edited by D. E. Matthews. VII, 385 pages. 1977.

Vol. 19: Measuring Selection in Natural Populations. Edited by F. B. Christiansen and T. M. Fenchel. XXXI, 564 pages. 1977.

Vol. 20: J. M. Cushing, Integrodifferential Equations and Delay Models in Population Dynamics. VI, 196 pages. 1977.

Vol. 21: Theoretical Approaches to Complex Systems. Proceedings 1977. Edited by R. Heim and G. Palm. VI, 244 pages. 1978.

Vol. 22: F. M. Scudo and J. R. Ziegler, The Golden Age of Theoretical Ecology: 1923–1940. XII, 490 pages. 1978.

Vol. 23: Geometrical Probability and Biological Structures: Buffon's 200th Anniversary. Proceedings 1977. Edited by R. E. Miles and J. Serra. XII, 338 pages. 1978.

Vol. 24: F. L. Bookstein, The Measurement of Biological Shape and Shape Change. VIII, 191 pages. 1978.

Vol. 25: P. Yodzis, Competition for Space and the Structure of Ecological Communities. VI, 191 pages. 1978.

Lecture Notes in Biomathematics

Managing Editor: S. Levin

47

Eckart Frehland

Stochastic Transport Processes in Discrete Biological Systems

Springer-Verlag
Berlin Heidelberg New York 1982

Author

Eckart Frehland
Fakultäten für Biologie und Physik, Universität Konstanz
7750 Konstanz, Federal Republic of Germany

AMS Subject Classifications (1980): 60, 62, 82, 92

ISBN-13: 978-3-540-11964-7 e-ISBN-13: 978-3-642-47511-5
DOI: 10.1007/978-3-642-47511-5

2141/3140-543210

Preface

These notes are in part based on a course for advanced students in the applications of stochastic processes held in 1978 at the University of Konstanz. These notes contain the results of recent studies on the stochastic description of ion transport through biological membranes. In particular, they serve as an introduction to an unified theory of fluctuations in complex biological transport systems. We emphasize that the subject of this volume is not to introduce the mathematics of stochastic processes but to present a field of theoretical biophysics in which stochastic methods are important.

In the last years the study of membrane noise has become an important method in biophysics. Valuable information on the ion transport mechanisms in membranes can be obtained from noise analysis. A number of different processes such as the opening and closing of ion channels have been shown to be sources of the measured current or voltage fluctuations. Biological transport systems can be complex. For example, the transport process can be coupled to other processes such as chemical reactions and take place in discontinuous structures of molecular dimensions. Furthermore, since there are strong electric fields or high concentration gradients across biological membranes ion transport processes of biological relevance are mostly processes far from equilibrium. For these reasons the development of new theoretical concepts has been necessary. The concept of transport in discrete systems has turned out to be more appropriate than continuum models.

There are two purposes for these notes: the first is to familiarize the reader with the theoretical background which is necessary for a satisfactory analysis of electrical noise in ion transport through biological membranes. Hence a number of applications of the general concepts to special transport models are discussed; the second is to show that the developed concepts may also be applied for the investigation of general properties of systems in nonequilibrium states. A main result of this study is the invalidity of the fluctuation-dissipation theorem for transport fluctuations at nonequilibrium steady states as consequence of the vectorial character of the considered (electrical) transport processes, while for scalar quantities the fluctuation dissipation theorem can be extended to non-equilibrium states.

I thank G.Adam, H.A.Kolb, P.Läuger and W.Stephan for many helpful discussions during the preparation of these notes. A number of numerical calculations have been done with the help of K.H.Faulhaber and B.Kleutsch at the Rechenzentrum of the University of Konstanz. Financial support (Heisenberg fellowship) by the Deutsche Forschungsgemeinschaft is gratefully acknowledged.

E.Frehland

Contents

A. Stochastic Processes

Contrariwise to a deterministic process governed by deterministic laws a stochastic or random process is controlled by probabilistic laws. Consider an experimental set-up by which an observable quantity or a set of observable quantities X(t) as a function of time can be measured: If the time course of X(t) is governed by probabilistic laws, different measurements of X(t) with the same experimental set-up at different times or with identical set-ups at the same time may yield different results X(t), representing different possible realizations of X(t). The ensemble of all possible realizations of X(t) is a stochastic (random) process. X(t) is called a random variable, or a set of random variables, respectively. Stochastic processes may be discrete or continuous in X and in time t. We will mainly consider processes which are discrete in X (e.g. integers) and continuous in time.

There is a large literature dealing with the mathematical fundamental theory of stochastic processes (e.g. Doob, 1953, Karlin, 1965). However, it is possible to get a knowledge which is sufficient for a great variety of applications in natural sciences, by adopting a less rigorous and more heuristic approach, as has been presented e.g. in the textbook by Bailey(1964). We mainly restrict this introductory part to a summarization of some definitions and facts, the knowledge of which is necessary for those readers who want to follow this volume without further study of mathematical textbooks.

1. Expectation Values, Moments, Variance, Correlations

Of central importance for the theoretical and experimental analysis of stochastic processes are averaged quantities. Though in most cases the experimentalist measures time averages, the theoretical approach is based on ensemble averaged quantities. If the stochastic process is ergodic (see below), ensemble averages and the corresponding time averages are equal. An ensemble average of some function f of X(t) is got by averaging over all possible realizations of X at time t. It is called the expectation value of f and denoted by $\langle f \rangle$:

$$\langle f(X(t)) \rangle \quad : \text{Expectation value of f at time t} \tag{A.1.1}$$

Moments

The most simple ensemble average is the expectation or mean value of X(t) itself. It is called the first moment of X:

$$\langle X(t) \rangle \quad : \text{Expectation value of X, first moment} \tag{A.1.2}$$

The mean square value of X is called the second moment:

$$\langle X^2(t) \rangle \quad : \text{Mean square value, second moment} \tag{A.1.3}$$

Generally:

$$\langle X^k(t) \rangle \quad : \text{k-th moment of X} \tag{A.1.4}$$

$$k = 1, 2, 3 \ldots$$

Often one is especially interested in the averaged deviations of X(t) from the mean value $\langle X(t) \rangle$ or the magnitude of the fluctuations around $\langle X(t) \rangle$. For this reason one introduces the so-called central moments or cumulants. Of special importance is the second central moment or variance:

$$\sigma^2(t) = \langle (X(t) - \langle X(t) \rangle)^2 \rangle \tag{A.1.5}$$

Generally the k-th central moment or cumulant is:

$$\sigma^k(t) = \langle (X(t) - \langle X(t) \rangle)^k \rangle \qquad (A.1.6)$$

The k-th central moment can always be expressed by the lower order moments. Explicitely for the variance holds the important relation:

$$\langle (X(t) - \langle X(t) \rangle)^2 \rangle = \langle X^2(t) \rangle - \langle X(t) \rangle^2 \qquad (A.1.7)$$

The validity of this relation may simply be shown by using the fact that the expectation value of a sum of quantities is equal to the sum of the expectation values of these quantities.

Correlations

Up to now we have considered quantities averaged at one time t. Further important information about the mechanisms underlying a stochastic process may be contained in the correlations between the measured quantities at different times. If X(t) represents a single variable, the most important correlation is the autocorrelation function C(t,s), by which an ensemble averaged correlation between X at time t and X at time (t+s) is defined:

$$C(t,s) = \langle X(t) \quad X(t+s) \rangle \qquad (A.1.8)$$

Often the autocorrelation function $C_{\Delta X}(t,s)$ of the deviation

$$\Delta X(t) = X(t) - \langle X(t) \rangle \qquad (A.1.9)$$

from the expectation value $\langle X(t) \rangle$ is used, namely

$$C_{\Delta X}(t,s) = \langle \Delta X(t) \quad \Delta X(t+s) \rangle \qquad (A1.10)$$

For s=0 C(t,o) is the second moment and $C_{\Delta X}(t,o)$ the variance of X(t):

$$C(t,o) = \langle X^2 \rangle \tag{A.1.11}$$

$$C_{\Delta X}(t,o) = \sigma^2(t)$$

Information about the correlations between different quantities Y and Z may be obtained by the crosscorrelation $C_{YZ}(t,s)$:

$$C_{YZ}(t,s) = \langle Y(t) \quad Z(t+s) \rangle \tag{A.1.12}$$

The crosscorrelation $C_{YZ}(t,s)$ may be different from the crosscorrelation $C_{ZY}(t,s)$. Further information may be contained in higher order correlations which are obtained by comparison of quantities at more than two different times.

2. Probabilities

We suppose that the reader has an intuitive feeling for the idea of probability: In the experiment of throwing a die the probability of getting a special number is 1/6. Probabilities are nonnegative and not greater than one.

If the possible values of a random variable X are discrete and denoted by X_i, the probability that X assumes the special value X_i is $P(X_i)$. Clearly,

$$\sum_{\text{all } i} P(X_i) = 1, \tag{A.2.1}$$

expressing the certainty that in one experiment X takes exactly one value X_i.

If the random variable assumes a continuous range of values, than we define the probability density p(X), where p(X) dX is the probability of finding the value of the random variable in the interval between X and X+dX. p(X) satisfies the relation

$$\int p(X)\, dX = 1 \tag{A.2.2}$$

Throughout this article probabilities are denoted by P and probability densities by p. In most cases we shall assume discrete processes and use probabilities P. For a continuous process the corresponding relations are found by replacing P(X) by p(X) dX and sums by integrals. The probability distribution function D(X) denotes the probability that the random variable is smaller than X:

$$D(X_1) = P(X \leq X_1) \tag{A.2.3}$$

Obviously, for continuous X the probability density function p(x) is related to the distribution function d(x) through

$$p(X) = \frac{dd(X)}{dX} \tag{A2.4}$$

2.1 Expectation Values and Probabilities

The expectation values may be defined with the use of the probabilities. If we admit time dependence, the expectation value $\langle X(t)\rangle$ in discrete processes is given from the probability P(X,t) by

$$\langle X(t)\rangle = \sum_{\text{all } X} X\, P(X,t) \tag{A.2.5}$$

The summation has to be taken over all possible values X. Note that the time dependence of $\langle X(t)\rangle$ comes in by the time dependence of P(X,t). Correspondingly the moments $\langle X^k(t)\rangle$ are

$$\langle X^k(t)\rangle = \sum_{\text{all } X} X^k P(X,t) \tag{A.2.6}$$

Generally, the expectation value $\langle f(X(t))\rangle$ is

$$\langle f(X(t))\rangle = \sum_{\text{all } X} f(X)\, P(X,t) \tag{A.2.7}$$

Conditional and Joint Probabilities

Often, in time dependent processes one needs the probability that X is the value of the random variable at time t under the condition that the value was X' at time t'. This conditional probability is denoted by

$$P(X,t/X', t').$$

It must be distinguished from the joint probability

$$P(X,t;\ X',t'),$$

which is the probability that at time t the value is X and at time t' the value is X'. Between conditional and joint probability holds the relation which is clear from the definitions:

$$P(X,t;\ X',t') = P(X,t/X',t')\ P(X',t') \tag{A.2.8}$$

Correspondingly, also higher order probabilities may be defined.

The autocorrelation function C(t,s) may be expressed by the joint probability:

$$C(t,s) = \langle X(t)\ X(t+s)\rangle = \sum_{X,X'} X\cdot X'\cdot P(X,t;X',t+s) \qquad (A.2.9)$$

Conditionally Averaged Values

The introduction of the conditional probability makes possible the definition of conditional averages. Often one is interested in the expectation value of X at time t under the (initial) condition that it was X' at time t'. It is given by

$$\langle X(t)\rangle_{(X',t')} = \sum_{X} X\ P(X,t/X',t') \qquad (A.2.10)$$

Correspondingly, conditional averages for other quantities may be defined. These subensemble averages are important for Markov processes (see part B) where the time course of probabilities is determined by only one initial condition. In case X stands for a set of variables, the probabilities and averages are defined analogously.

3.Binomial, Poisson, Normal Distrubutions

Binomial Distribution

Consider a special experiment, where the probability for a special (positive) result is P. Then the probability for another (negative) result is clearly (1-P). We ask for the probability that in m identical and independent experiments n have positive results. This probability $P_m(n)$ is given by the binomial law

$$P_m(n) = \binom{m}{n} P^n (1-P)^{m-n} \qquad (A.3.1)$$

with

$$\binom{m}{n} = \frac{m!}{n!\,(m-n)!}$$

In (A.3.1) the possible values n of the random variable are positive integers. $P_m(n)$ satisfies relation (A.2.1):

$$\sum_{n=0}^{m} P_m(n) = 1 \qquad (A.3.2)$$

The summation in (A.3.2) is taken over all possible values n. With the use of relations (A.2.5)-(A.2.6) the expectation values can be calculated from (A.3.1). The results for the first two moments $\langle n \rangle$, $\langle n^2 \rangle$ and the variance σ^2 are

$$\begin{aligned} \langle n \rangle &= m\,P \\ \langle n^2 \rangle &= m^2P^2 + mP(1-P) \qquad (A.3.3) \\ \sigma^2 &= m\,P(1-P) = \langle n \rangle\,(1-P) \end{aligned}$$

Poisson Distribution

The Poisson distribution is derived from the binomial law by the limiting process

$$\begin{aligned} &m \to \infty, \\ &P \to 0, \\ &mP = \langle n \rangle \text{ finite}, \qquad (A.3.4) \end{aligned}$$

where m and P remain the number of identical independent experiments and the probability of a positive result of a single experiment, respectively. The probability P(n) that n experiments have positive results is given by the Poisson law:

$$P(n) = \frac{\langle n \rangle^n}{n!} e^{-\langle n \rangle} \qquad (A.3.5)$$

(A.3.5) is used instead of (A.3.1) in cases of a sufficiently great number of experiments. P(n) satisfies relation (A.2.1):

$$\sum_{n=0}^{\infty} P(n) = 1 \qquad (A.3.6)$$

Second moment and variance can be directly derived from (A.3.3)

$$\langle n^2 \rangle = \langle n \rangle^2 + \langle n \rangle \qquad (A.3.7)$$
$$\sigma^2 = \langle n \rangle$$

Normal Distribution

A random variable X is said to follow a normal distribution if its probability density p(X) is given by:

$$p(x) = \frac{1}{\sqrt{2\pi\sigma^2}} \exp\left[- \frac{(X - \langle X \rangle)^2}{2\sigma^2} \right] \qquad (A.3.8)$$

$\langle X \rangle$: expectation (mean) value, σ^2: the variance.

The importance of the normal distribution is a consequence of the central-limit-theorem which, roughly speaking, asserts that the normal distribution will result in general, when a large number of independent random variables are summed to obtain one new random variable.

4. Stationarity and Ergodicity

Stationarity

A stochastic process is called stationary, if $P(X,t)$ is independent of time:

$$P(X,t) = P(X) \qquad (A.4.1)$$

Otherwise it is called nonstationary. As consequence of (A.4.1) all moments are time-independent. Often a stochastic process is called (weakly) stationary, if the first two moments are time-independent.

The autocorrelation function $C(t,s)$ in stationary processes is independent of time t and satisfies the following relations, which may easily be derived from the definition (A.1.8) of $C(t,s)$ and the definition of stationarity:

$$\begin{aligned} C(t,s) &= C(0,s) = C(s) && \text{a)} \\ C(s) &= C(-s) = C(|s|) && \text{b)} \qquad (A.4.2) \\ |C(s)| &\leq C(o) \geq 0 && \text{c)} \end{aligned}$$

Ergodicity

Of great practical importance is the special class of ergodic processes. A stationary stochastic process is called ergodic, if the ensemble averages (expectation values) are equal to the corresponding time averages over one realization $X_i(t)$ of the stochastic process:

$$\langle f(X) \rangle = \overline{f(X(t))}$$

$$\bar{f} := \lim_{T\to\infty} \frac{1}{2T} \int_{-T}^{+T} f(X\ (t))\, dt \qquad (A.4.3)$$

Time averages in this article are denoted by $\overline{f}$. Especially, the moments are in ergodic processes:

$$\langle X \rangle = \overline{X_i}$$

$$\langle X^k \rangle = \overline{X_i^k} \qquad (A.4.4)$$

$$\sigma^2 = \overline{(X_i - \overline{X}_i)^2}$$

and the autocorrelation function

$$C(s) = \overline{X_i(t)\ X_i(t+s)} \qquad (A.4.5)$$

We note that ergodicity necessarily implies stationarity.
In experimental investigations of stationary processes mostly ergodicity is assumed. Fortunately this has been successful in most cases. Ergodic processes can be investigated with <u>one</u> experimental set-up (one realization of the stochastic process) and the quantities characterizing the process can be measured by time averaging.

Nevertheless, we must keep in mind that stationarity not necessarily implies ergodicity, e.g. stationary processes may be nonergodic if the realizations $X_i(t)$ depend in some random way on starting conditions. A simple example is the stationary process defined by the realizations

$$X_i(t) = A(i) \sin\left[\omega t + \theta(i)\right] \qquad (A.4.6)$$

where amplitude $A(i)$ and phase $\theta(i)$ are random variables.

A simple classification of stochastic processes is given by the following scheme:

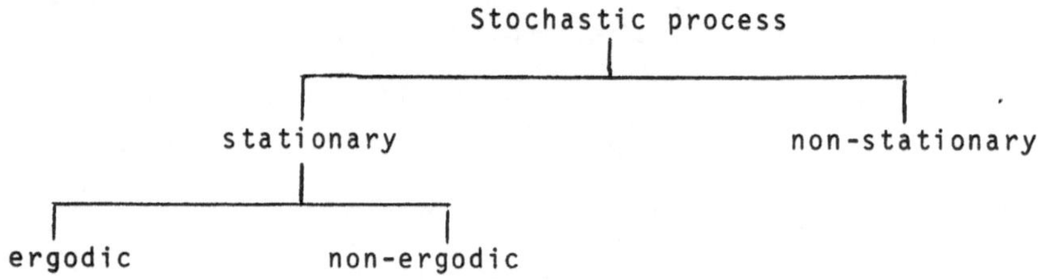

B. Analysis of Stationary Stochastic Processes

1. Basic Concepts of Noise Analysis

In this section we describe basic concepts of noise analysis of stationary processes. The contents and further details may be found in most of the standard books on noise analysis, e.g. Bendat, Piersol (1971), Bell (1960), Bittel, Storm (1971), Pfeifer (1959), van der Ziel (1970, 1976).

The autocorrelation function C(s) and the spectral density or power spectrum G(f) are the basic quantities used in noise analysis. Though the theoretical definition of both quantities is done by ensemble averages, the experimental measurement of C(s) and G(f) in most cases can be performed only by time averaging over a single realization X(t) under the assumption that the stationary process is even ergodic.

1.1 Spectral Density

The spectral density describes the frequency composition of stationary stochastic data by the spectral decomposition of the mean square value. In ergodic processes the spectral density or power spectrum of a quantity X(t) (e.g. current or voltage) is measured as follows (see Fig.B1): X(t) is decomposed into its spectral parts by filtering with a band pass filter having sharp characteristics. Square averaging of these sinusoidal signals yields the spectral density G(f) as spectral decomposition of $\langle x^2 \rangle$:

$$\langle x^2 \rangle = \int_0^\infty G(f)\, df \qquad (B.1.1)$$

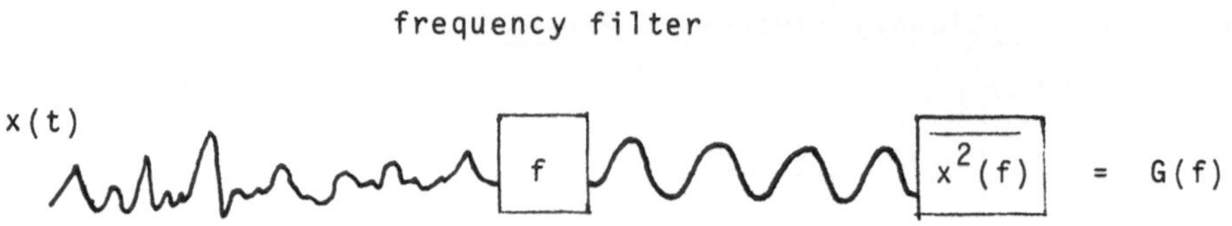

Fig. B1: Measurement of spectral density in ergodic processes

The mathematical definition of the spectral density G(f) of stationary processes is done as follows: take a Fourier expansion of some realization X(t) in the interval $-T \le t \le +T$:

$$X(t) = \int_{-\infty}^{+\infty} a_T(f)\, e^{i\omega t}\, df, \quad \omega = 2\pi f,$$

$$a_T = \int_{-T}^{+T} X(t)\, e^{-i\omega t}\, dt \qquad (B.1.2)$$

The expansion into positive frequencies is:

$$X(t) = \int_{0}^{\infty} (a_T\, e^{i\omega t} + a_T^{*}\, e^{-i\omega t})\, df$$

with a_T^{*} = complex conjugate of a_T, or

$$X(t) = \int_{0}^{\infty} 2|a_T| \cos(\omega t + \varphi_\omega)\, df \qquad (B.1.3)$$

($|a_T|$: absolute value of a_T and a_T^{+}, φ_ω: frequency dependent phase). Square averaging, ensemble averaging and taking the limit as $T \to \infty$ yields the definition of G(f):

$$G(f) := \lim_{T\to\infty} \frac{1}{2T}\, 4 \left\langle |a_T{}^2| \right\rangle =$$

$$= \lim_{T\to\infty} \frac{1}{2T}\, 4 \left\langle a_T\, a_T^{*} \right\rangle, \quad f \ge 0 \qquad (B.1.4)$$

For ergodic processes the ensemble average in (B.1.4) may be omitted. G(f) is defined for positive frequencies.

There may arise some confusion, because often a spectral density S(f) is mathematically defined for positive and negative frequencies by:

$$S(f) = \lim_{T\to\infty} \frac{1}{2T} \left\langle a_T \, a_T^+ \right\rangle , \qquad (B.1.5)$$

$$-\infty \leq f \leq +\infty$$

Throughout this article we shall use G(f) as in definition (B.1.4) By comparison we find G(f) = 4 S(f) for f ≥ o (B.1.6)

1.2 Autocorrelation Function and Wiener-Khintchine Relations

The autocorrelation function C(s) according to (A.1.9), and the spectral density G(f) are measurable quantities which may be used for the analysis of stochastic processes: C(s) contains information in the time domain, G(f) in the frequency domain. Mathematically, both types of information are equivalent because according to the Wiener-Khintchine relations C(s) and G(f) can be calculated from each other by Fourier transformation (Wiener, 1930, Khintchine, 1934):

$$G(f) = 4 \int_0^\infty C(s) \cos \omega s \; ds \qquad a)$$

(B.1.7)

$$C(s) = \int_0^\infty G(f) \cos \omega s \; df \qquad b)$$

Naturally an essential condition for the validity of these relations is that the integral on the right-hand side of (B.1.7)a exist for all ω, i.e. in the limit $s \to \infty$, C(s) must vanish sufficiently fast. Physically, for sufficiently long times s, X(t) and X(t+s) become uncorrelated.

Though C(s) and G(f) contain equivalent information, this information is contained in different form, where one form is often more desirable than the other for specific applications.

2. Poisson Processes, Carson's Theorem

A special class of stationary stochastic processes important for many practical applications, are the so-called Poisson processes:

> A stationary process is called a Poisson process, if the realizations $X(t)$ consist of a sequence of identical events $g(t)$ occurring independently of each other with a mean number per unit time (average frequency) λ :
>
> $$X(t) = \sum_i g(t-t_i) \qquad \text{(B.2.1)}$$

Obviously the Poisson process is ergodic. In many cases the single events (pulses) are assumed to start at t_i, i.e.

$$g(t-t_i) = 0 \quad \text{for} \quad t-t_i < 0 \qquad \text{(B.2.2)}$$

Since the events occur independently, the probability $P(n, \tau)$, that within a time interval τ, n pulses begin , is given by the Poisson law (A.3.5)

$$P(n, \tau) = \frac{(\lambda \cdot \tau)^n}{n!} e^{-\lambda\tau} \qquad \text{(B.2.3)}$$

$\lambda \cdot \tau$ is the mean (expectation) value of the number of pulses starting within time τ. Because all events occur independently, τ may be decomposed into small time intervals $\Delta\tau$. In the limit $\Delta\tau \to 0$ the conditions for the validity of the Poisson law are satisfied with $m = \tau/\Delta\tau$ as the number of independent experiments ,and $p = \lambda\Delta\tau$ as the probability that within $\Delta\tau$ an event is starting.

In the following we give, without derivation, results from the theory of Poisson processes and explicit formulae for some special examples of elementary events g(t). For mathematical derivation the reader is referred to the literature.

2.1. Variance, Autocorrelation Function, Spectral Density

In Poisson processes the mean value, variance, autocorrelation function and spectral density can directly be calculated from the time course of the single events $g(t-t_i)$. The mean (expectation) value $\langle X \rangle$ is simply

$$\langle X \rangle = \lambda \int_{-\infty}^{+\infty} g(t)\, dt \qquad \text{(B.2.4)}$$

The autocorrelation function C(s) is given by the convolution integral

$$C(s) = \lambda \int_{-\infty}^{+\infty} g(t)\, g(t+s)\, dt + \langle X \rangle^2 \qquad \text{(B.2.5)}$$

Often only the autocorrelation function $C_{\Delta X}(s)$ of the fluctuating part of X is considered:

$$C_{\Delta X}(s) = \lambda \int_{-\infty}^{+\infty} g(t)\, g(t+s)\, dt \qquad \text{(B.2.6)}$$

Hence the variance σ^2 is

$$\sigma^2 = C_{\Delta X}(o) = \lambda \int_{-\infty}^{+\infty} g(t)^2\, dt \qquad \text{(B.2.7)}$$

The relations (B.2.4) and (B.2.7) together are known as Campbell's theorem.

The spectral density $G(\omega)$ is determined by the Fourier transform of the single event

$$\psi_g(\omega) = \int_{-\infty}^{+\infty} g(t)\, e^{-i\omega t}\, dt \qquad \text{(B.2.8)}$$

Application of the Wiener-Khintchine relations (B.1.7) to (B.2.5) yields

$$G(\omega) = 2\lambda |\psi_g(\omega)|^2 + \langle X \rangle^2 \delta(\omega) \qquad \text{(B.2.9)}$$

$|\psi_g(\omega)|$: absolute value of $\psi_g(\omega)$. The spectral density $G_{\Delta X}(\omega)$ of the fluctuating part is

$$G_{\Delta X}(\omega) = 2\lambda |\psi_g(\omega)|^2 \qquad \text{(B.2.10)}$$

(B.2.9), (B.2.10) are known as Carson's theorem.

2.2 Special Examples

We now present the autocorrelation function and spectral density for some special cases of pulses $g(t)$.

Rectangular Events

If a Poisson process is generated by rectangular pulses with height h, duration τ and rate λ (see Fig. B2a) the mean value $\langle X \rangle$ and autocorrelation function $C_{\Delta X}(s)$ are simply as in (B.2.4) and (B.2.6) (for a simple derivation see e.g. Bittel, Storm (1971):

$$\langle X \rangle = \lambda \cdot h \cdot \tau \qquad \text{(B.2.11a)}$$

$$C_{\Delta X}(s) = \begin{cases} \lambda\, h^2(\tau - |s|) & \text{for} \quad |s| < \tau \\ 0 & \text{for} \quad |s| > \tau \end{cases} \qquad \text{(B.2.11b)}$$

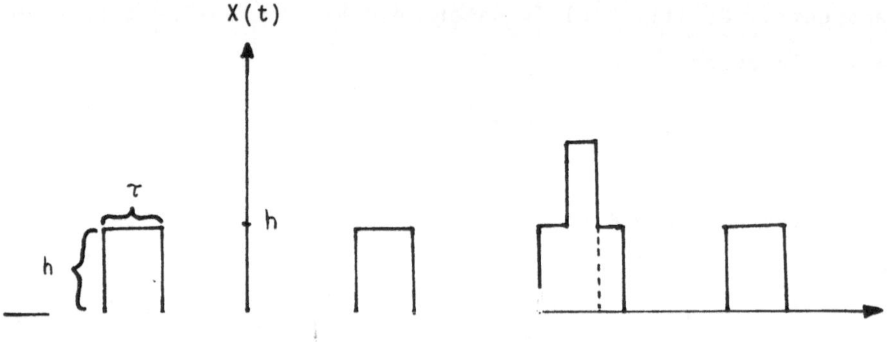

Fig. B2a: Poisson process built up by rectangular events

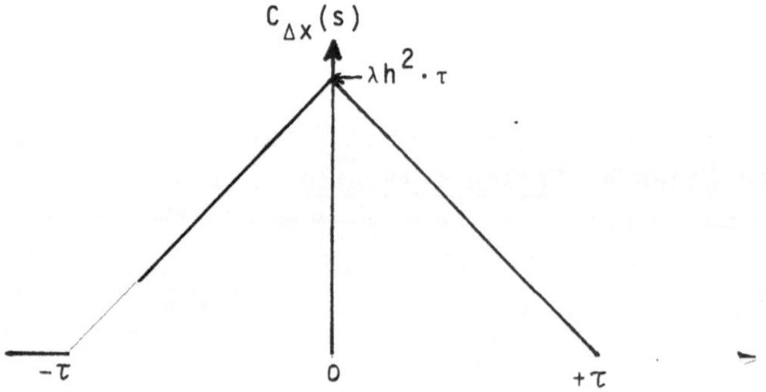

Fig. B2b: Autocorrelation function $C_{\Delta x}(s)$ for a Poisson process built up by rectangular events according to Fig. B1.

δ-pulses

From rectangular pulses we obtain δ-pulses in the limit $h \to \infty$, $\tau \to 0$ with

$$\tau \cdot h = H \tag{B.2.12}$$

Then from (B.2.11) and (B.1.7) one gets mean value, autocorrelation function and spectral density:

$$\langle X \rangle = \lambda \cdot H \tag{B.2.13a}$$

$$C_{\Delta X}(s) = \lambda\ H^2\ \delta(s) \tag{B.2.13b}$$

$$G_{\Delta X}(\omega) = 2\lambda \cdot H^2 \tag{B.2.13c}$$

In a similar way by taking the limit from rectangular events to δ-pulses we shall give below in part C (section 4) a mathematical foundation of the general relation (C.4.2) describing transport fluctuations in discrete systems.

2.3 Ions Crossing Membranes: Shot Noise

In a very simple model of ion transport through membranes we can consider the membrane, i.e. the boundary layer between two aqueous solutions, as a single potential barrier over which the ions have to jump.

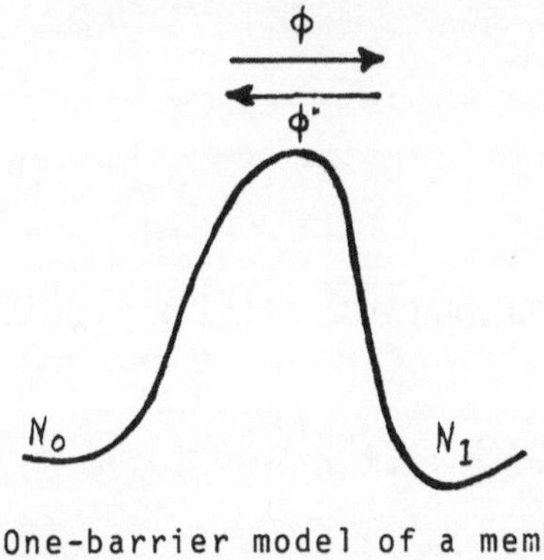

Fig. B3: One-barrier model of a membrane.

In Fig. B3 the ionic concentrations on the left- and right-hand sides of the membranes are indicated by the indices 0, 1 respectively, and regarded as constant reservoirs. The duration of the ionic jumps over the potential barrier is assumed to be very short. Hence the unidirectional fluxes $\phi'(\phi'')$ from the left (right) to the right (left) can be treated as δ-pulse sequences.

Under the assumption that interactions between the ions can be neglected, the mean (expectation) values $\langle\phi'\rangle, \langle\phi''\rangle$ of the fluxes are time independent. They represent the mean rates of ionic jumps in the two directions and are given by

$$\langle\phi'\rangle = k' N_0$$
$$\langle\phi''\rangle = k'' N_1 \quad , \tag{B.2.14}$$

k', k" being rate constants for ionic jumps to the right, left respectively.

Therefore the whole transport process is a superposition of two independent Poisson processes built up by δ-pulses. With (B.2.13c) and H=1 the spectral densities $G_{\Delta\phi'}(\omega)$ and $G_{\Delta\phi''}(\omega)$ of the unidirectional fluxes are

$$G_{\Delta\phi'}(\omega) = 2\langle\phi'\rangle = 2\, k'\, N_0$$
$$G_{\Delta\phi''}(\omega) = 2\langle\phi''\rangle = 2\, k''\, N_1 \tag{B.2.15}$$

Introducing the net flux

$$\phi = \phi' - \phi'' \tag{B.2.16}$$

over the membrane, the corresponding spectral density $G_{\Delta\phi}(\omega)$ is

$$G_{\Delta\phi}(\omega) = G_{\Delta\phi'}(\omega) + G_{\Delta\phi''}(\omega) = 2(\langle\phi'\rangle + \langle\phi''\rangle) \tag{B.2.17}$$

In case a high electric voltage is applied to the membrane so that $\langle\phi''\rangle$ can be neglected, we get

$$G_{\Delta\phi}(\omega) = G_{\Delta\phi'}(\omega) = 2\langle\phi'\rangle \tag{B.2.18}$$

For application to experimental measurements of electric fluctuations we introduce the electric current J

$$J = ze_0\phi \qquad (B.2.19)$$

(z: valency; e_0: elementary electric charge).

In this case H in (B.2.13) is ze_0 and the spectral density $G_{\Delta J}(\omega)$ of current fluctuations becomes

$$G_{\Delta J}(\omega) = 2\, z^2 e_0^2\, (\langle \phi' \rangle + \langle \phi'' \rangle) \qquad (B.2.20)$$

For high applied voltage:

$$G_{\Delta J}(\omega) = 2\, ze_0\, J^s \qquad (B.2.21)$$

with the stationary current

$$J^s = ze_0\langle \phi' \rangle$$

(B.2.21) is the usual form of the well-known shot noise relation or Schottky's theorem (Schottky, 1918), which always holds if the current J consists of a sequence of independent pulses all carrying the same charge $q = ze_0$.

In part C, section 8, we will again treat the problem of shot noise under very general conditions and discuss the influence of interactions.

3. The Master Equation Approach to Fluctuations Around Steady States

Central part of this section will be the introduction into the so-called master equation approach to fluctuations in stationary Markow processes. A most profound description of this approach has been given by van Vliet and Fassett (1965) with applications to fluctuations in two- and multilevel semiconductors. Generally we will restrict to the presentation of definitions and results. Information about Markov-processes may be found in the standard mathematical literature, e.g. Bailey (1964), Bharucha-Reid (1960), Feller (1950). The master equation approach has been described e.g. by van Vliet and Fassett (1965), Lax (1960) and applied to membrane transport by Chen (1978).

In section 3.6 we briefly outline the treatment of fluctuations, if one starts directly from the solution of the master equation instead of the linearized phenomenological equations.
Apart from the avoidance of problems concerning e.g. the linearization properties of the phenomenological equations, it turns out that in this case, the structures of variance, correlation and noise spectrum matrices are drastically simplified. These results can be applied e.g. in the treatment of channel noise (section 3.7).

The following notations and conventions in matrix calculus will be used: Matrices are characterized by $\sim$.
The elements of a matrix are indicated by sub-indices i, j, k ..., e.g. a (nxm)-matrix $\underset{\sim}{A}$ has the elements A_{ik} ($i = 1, 2, \ldots, n$, $k = 1, 2, \ldots, m$), a column matrix $\underset{\sim}{N}$ is a (nxl)-matrix with elements N_i ($i = 1, 2, \ldots n$). The transpose of a matrix is indicated by $\tilde{}$, e.g. $\tilde{\underset{\sim}{A}}$ with the elements

$$\tilde{A}_{ik} = A_{ki} \tag{B.3.1}$$

The transpose $\tilde{N}$ of a column matrix is a row. Matrix multiplication $A \cdot B$ is defined by

$$(A \cdot B)_{ik} = \sum_{j=1}^{m} A_{ij} B_{jk} \qquad \text{(B.3.2a)}$$

(A: (nxm)-matrix, B: (mxp)-matrix), and

$$(A \cdot N)_i = \sum_{j=1}^{m} A_{ij} N_j \qquad \text{(B.3.2.b)}$$

(N: (mx1)-column matrix).

The inverse of a square (mxm)-matrix M, in case it exists, is M^{-1} with the components

$$M_{ik}^{-1} = \frac{(\text{adj } M)_{ik}}{\det M} \qquad \text{(B.3.3)}$$

(adj M: adjugate of M, det M: determinant of M).

3.1 Markov Processes

Consider a stochastic process presented by $N(t)$, where $N(t)$ is a set of n random variables $N_i(t)$, ,,, ,$N_n(t)$. This set of variables is assumed to be Markovian:

The stochastic process represented by the set of variables $N(t)$ is called a Markov Process, if a specification of state N' at time t' is sufficient to make prediction of the state $N(t)$ at a later time t. How the state N' is reached has no influence on the further probabilistic development of N.

We emphasize that in many cases the observable stochastic variables are not Markovian, but they depend uniquely on Markovian variables. For instance, in kinetic nerve channel models (e.g. the Hodgkin-Huxley model) with one open and several closed states, the set of variables representing the different channel states is a Markovian set of variables, while the number of channels in the open state, which determines the measurable conductivity, is not a Markovian set.

In the following we assume the Markov processes to be homogeneous in time. A Markov process is <u>homogeneous in time,</u> if a shift in time for the conditional probability $P(\underset{\sim}{N}, t|\underset{\sim}{N}', t')$ is irrelevant:

$$P(\underset{\sim}{N}, t|\underset{\sim}{N}', t') = P(\underset{\sim}{N}, t-t'|\underset{\sim}{N}', 0) \qquad (B.3.4)$$

Remember that the conditional probability $P(\underset{\sim}{N}, t|\underset{\sim}{N}', t')$ is defined as the probability of finding the system in state $\underset{\sim}{N}$, at time t under the condition that at time t' it was in state $\underset{\sim}{N}'$ (c.f. section A.2.2). Obviously a Markov process is homogeneous in time, if all external parameters controlling the system are constant in time.

3.2 Master Equation and Phenomenological Equations

The master equation describes the change in time of the conditional probability, expressed by its derivative with respect to time t. The general form of the master equation is

$$\frac{\partial P(\underset{\sim}{N}, t|\underset{\sim}{N}', 0)}{\partial t} = + \text{(sum, taking into account possible transitions to state } \underset{\sim}{N}) - \text{(sum, taking into account possible transitions away from state } \underset{\sim}{N}) \qquad (B.3.5)$$

In kinetic systems the small-time transition probabilities appearing on the right-hand side of (B.3.5) can easily be expressed by the transition rates per unit time $Q(\underset{\sim}{N};\underset{\sim}{N}')$ from $\underset{\sim}{N}'$ to $\underset{\sim}{N}$ for $\underset{\sim}{N}' \neq \underset{\sim}{N}$. $Q(\underset{\sim}{N};\underset{\sim}{N})$ is defined to be zero. Then the master equation takes a form, which will be used in the following:

$$\frac{\partial P(\underset{\sim}{N}, t|\underset{\sim}{N}', 0)}{\partial t} = \sum_{\text{all } \underset{\sim}{N}''} P(\underset{\sim}{N}'', t|\underset{\sim}{N}', 0) . Q(\underset{\sim}{N}, \underset{\sim}{N}'')$$

$$- \sum_{\text{all } \underset{\sim}{N}''} P(\underset{\sim}{N}, t|\underset{\sim}{N}', 0) \cdot Q(\underset{\sim}{N}''; \underset{\sim}{N}) \qquad \text{(B.3.6)}$$

The sums in (B.3.6) over $\underset{\sim}{N}''$ are to be taken over all possible states, which are assumed to be <u>discrete.</u> For processes, which are continuous in $\underset{\sim}{N}$, P is replaced by the probability density p and the sums by integrals. For a finite total number n of different states $\underset{\sim}{N}$ of the system we can write (B.3.6) in an alternative form (suppressing the initial condition), which will be used mainly in part C:

$$\frac{dP_\mu}{dt} = \sum_{\nu=1}^{n} M_{\mu\nu} P_\nu \qquad \text{(B.3.7)}$$

with $$M_{\mu\mu} = - \sum_{\substack{\nu=1 \\ \nu\neq\mu}}^{n} M_{\nu\mu}$$

Each state ν can take the values 1 or 0 according as the system is in state ν or not.

Though the master equation (B.3.6) is a linear system of first order differential equations for the conditional probabilities P with constant coefficients, its solution may be complicated, especially in cases of a great number of different states.

The following results concerning the solutions of (B.3.7) are given without proof (c.f. Schnakenberg (1976), Aschinger (1975)): (B.3.7) has at least one stationary solution, which is even unique in case the graph belonging to (B.3.7) is connected. If at time t=0 hold (B.3.7) and the condition $0 \le P_\nu \le 1$ for all ν, then this condition is satisfied for all later times. The nonvanishing eigenvalues of $\underset{\sim}{M}$ have negative real parts, i.e. the stationary solution is asymptotically stable.

In many cases for the analysis of the stochastic fluctuations of the Markovian variables around steady states it is not required to solve the master equation. It is possible to derive the necessary results by analytical manipulations of (B.3.6) without knowing the solution. E.g. a differential equation for the conditionally averaged expected value of $\underset{\sim}{N}$ (c.f.(A.2.10))

$$\left\langle \underset{\sim}{N}(t) \right\rangle_{\underset{\sim}{N}(0)} = \sum_{\text{all } \underset{\sim}{N}} \underset{\sim}{N} \cdot P(\underset{\sim}{N}, t \mid \underset{\sim}{N}(0)) \qquad (B.3.8)$$

may be obtained by multiplying (B.3.6) with $\underset{\sim}{N}$ and summing over all $\underset{\sim}{N}$. The result is

$$\frac{\partial \left\langle \underset{\sim}{N}(t) \right\rangle_{\underset{\sim}{N}(0)}}{\partial t} = \sum_{\text{all } \underset{\sim}{N}''} \underset{\sim}{A}(\underset{\sim}{N}'')\, P(\underset{\sim}{N}'', t \mid \underset{\sim}{N}(0)) = \left\langle \underset{\sim}{A}(\underset{\sim}{N}) \right\rangle_{\underset{\sim}{N}(0)} \qquad (B.3.9)$$

where the components A_i of

$$\underset{\sim}{A}(\underset{\sim}{N}'') = \sum_{\text{all } \underset{\sim}{N}} (\underset{\sim}{N} - \underset{\sim}{N}'')\, Q(\underset{\sim}{N}; \underset{\sim}{N}'') \qquad (B.3.9a)$$

are called the first order Fokker-Planck moments.
Difficulties in solving equation (B.3.9) for the first moment and the corresponding equations for the higher moments and central moments (cumulants) especially in nonlinear processes will be discussed below in part D.

3.3 Linearized Phenomenological Equations and Fundamental Solutions

In the following we assume that the phenomenological equations (B.3.9) have a stable steady state solution $\langle \underset{\sim}{N}^s \rangle = \underset{\sim}{N}^s$.

Linearization of the equation for the expected value $\underset{\sim}{N}(t)_{\underset{\sim}{N}(0)}$ around $\underset{\sim}{N}^s$ yields the linear phenomenological equations

$$\frac{d \left\langle \underset{\sim}{N}(t) \right\rangle_{\underset{\sim}{N}(0)}}{dt} = \underset{\sim}{M} \left\langle \underset{\sim}{N}(t) \right\rangle_{\underset{\sim}{N}(0)} + \underset{\sim}{Y} \qquad \text{(B.3.10)}$$

The steady state $\underset{\sim}{N}^s$ is given by the equation

$$\underset{\sim}{M}\, \underset{\sim}{N}^s + \underset{\sim}{Y} = 0 \qquad \text{(B.3.11)}$$

which is solved by standard methods. The treatment of fluctuations is done with the use of the linearization (B.3.10), which is a consistent approximation in case the (calculated) fluctuations are small enough to remain within the region where the linearization is a good approximation of (B.3.9).

From (B.3.10) we get, for the deviations

$$\alpha_i = N_i - N_i^s , \qquad \text{(B.3.12)}$$

the homogeneous equations

$$\frac{d \left\langle \underset{\sim}{\alpha}(t) \right\rangle_{\underset{\sim}{\alpha}(0)}}{dt} = \underset{\sim}{M} \left\langle \underset{\sim}{\alpha}(t) \right\rangle_{\underset{\sim}{\alpha}(0)} \qquad \text{(B.3.13)}$$

The fundamental solutions

(B.3.10) and (B.3.13) can be solved by standard methods extensively described in the literature. Formally (B.3.13) is solved by the matrix exponential function

$$\bar{\underset{\sim}{\Omega}}(t) := \exp(-\underset{\sim}{M}t) = \sum_{k=0}^{\infty} (-1)^k (\underset{\sim}{M})^k / k! \qquad \text{(B.3.14)}$$

with

$$(\underset{\sim}{M})^2 = \underset{\sim}{M}\underset{\sim}{M}, \qquad \text{etc.}$$

With (B.3.14) the solution of (B.3.13) is

$$\langle \underset{\sim}{\alpha}(t) \rangle_{\underset{\sim}{\alpha}(0)} = \exp(-\underset{\sim}{M}t)\ \tilde{\underset{\sim}{\alpha}}(0) \qquad \text{(B.3.15)}$$

The matrix exponential $\underset{\sim}{\Omega}$ is often called the fundamental solution matrix (e.g. Braun, 1978). We use a fundamental solution matrix $\underset{\sim}{\Omega}(t)$ which is defined in a slightly different way: The elements $\Omega_{ik}(t)$ are the solutions

$$\Omega_{ik}(t) = \langle \alpha_i(t) \rangle \quad (\alpha_k(0) = 1,\ \alpha_j(0) = 0 \text{ otherwise}) \qquad \text{(B.3.16a)}$$

for $\underset{\sim}{Y} \neq 0$,

and

$$\Omega_{ik}(t) = \langle \alpha_i(t) \rangle \quad (\alpha_k(0) + N_k^s = 1,\ \alpha_j + N_j^s = 0 \text{ otherwise}) \qquad \text{(B.3.16b)}$$

for $\underset{\sim}{Y} = 0$.

With (B.3.16) for arbitrary initial condition the solution of (B.3.13) is obtained from $\underset{\sim}{\Omega}(t)$ by

$$\langle \alpha(t) \rangle_{\underset{\sim}{\alpha}(0)} = \underset{\sim}{\Omega}(t)\ \tilde{\underset{\sim}{\alpha}}(0) \qquad \text{for } \underset{\sim}{Y} \neq 0 \qquad \text{(B.3.17a)}$$

$$\left\langle \underset{\sim}{\alpha}(t) \right\rangle_{\underset{\sim}{\alpha}(0)} = \underset{\sim}{\Omega}(t)\ \underset{\sim}{\tilde{N}}(0) = \underset{\sim}{\Omega}(t)\ \underset{\sim}{\tilde{\alpha}}(0) \qquad \text{for } \underset{\sim}{Y} = 0 \qquad (B.3.17b)$$

The only difference between the definition of the matrix exponential $\bar{\underset{\sim}{\Omega}}(t)$ and $\underset{\sim}{\Omega}(t)$ is that, for $\underset{\sim}{Y} = 0$, we have subtracted from $\bar{\Omega}(t)$ the (nontrivial) steady state solution, because in this case $\bar{\underset{\sim}{\Omega}}(t)$ contains a nontrivial steady state solution $\lim_{t\to\infty} \bar{\underset{\sim}{\Omega}}(t) \neq 0$, while for $\underset{\sim}{Y} \neq 0$ $\lim_{t\to\infty} \underset{\sim}{\Omega}(t) = 0$. Hence for $\underset{\sim}{Y} = 0$ <u>and</u> $\underset{\sim}{Y} \neq 0$

$$\underset{\sim}{\Omega}(t) = \bar{\underset{\sim}{\Omega}}(t) - \lim_{t\to\infty} \bar{\underset{\sim}{\Omega}}(t) \qquad (B.3.18)$$

The distinction between the cases $\underset{\sim}{Y} = 0$ and $\underset{\sim}{Y} \neq 0$ is necessary because of the existence of a unique nontrivial steady state solution $\underset{\sim}{N}^s \neq 0$ of (B.3.11). It is known (e.g. Braun, 1978) that this means for $\underset{\sim}{Y} = 0$ the inclusion of the nontrivial solution $\underset{\sim}{N}^s$ in the matrix exponential. Systems with $\underset{\sim}{Y} = 0$ are often called 'closed' systems (see e.g. Chen, 1978). They admit conservation relations (the rank of $\underset{\sim}{M}$ is $< n$) and may be reduced to the case $Y \neq 0$ ('open' system) by elimination of variables with the use of these conservation relations.

We emphasize that for the master equation in the form (B.3.7) the fundamental solution matrix may be introduced in the same way as for the case $\underset{\sim}{Y} = 0$ in (B.3.13). This will be used below.

3.4 Variances, Correlations, Spectra

Variance matrix

In extension of the variance σ^2 in the one variable case, for multivariate problems the symmetric variance matrix $\underset{\sim}{\sigma}^2$ is introduced:

$$\underset{\sim}{\sigma}^2 = \langle \underset{\sim}{\alpha} \tilde{\underset{\sim}{\alpha}} \rangle$$

$$\sigma_{ik}^2 = \langle \alpha_i \alpha_k \rangle = \sigma_{ki}^2 \qquad \text{(B.3.19)}$$

An important relation for $\underset{\sim}{\sigma}^2$ can be derived from the master equation (B.3.6) similarly as the linearized phenomenological equations (B.3.10) by multiplication with $\underset{\sim}{\alpha} \tilde{\underset{\sim}{\alpha}}$, summation over all possible states, expansion about the steady state $\underset{\sim}{N}^s$ and taking the limit $t \to \infty$ (c.f. van Vliet and Fassett, 1965, Lax, 1960, Chen, 1978):

$$\underset{\sim}{\sigma}^2 \tilde{\underset{\sim}{M}} + \underset{\sim}{M} \underset{\sim}{\sigma}^2 = \underset{\sim}{B}(\underset{\sim}{N}^s) \qquad \text{(B.3.20)}$$

The components B_{ik} of the symmetric matrix $\underset{\sim}{B}(\underset{\sim}{N})$ are the so-called second order Fokker-Planck moments and defined by

$$\underset{\sim}{B}(\underset{\sim}{N}') = \sum_{\text{all } \underset{\sim}{N}} (\underset{\sim}{N}-\underset{\sim}{N}')(\tilde{\underset{\sim}{N}}-\tilde{\underset{\sim}{N}}')Q(\underset{\sim}{N};\underset{\sim}{N}') \qquad \text{(B.3.21)}$$

Hence the elements of $\underset{\sim}{B}(\underset{\sim}{N}^s)$ are

$$B_{ik}(\underset{\sim}{N}^s) = \sum_{\text{all } \underset{\sim}{\alpha}} \alpha_i \alpha_k \quad Q(\underset{\sim}{N};\underset{\sim}{N}') \qquad \text{(B.3.22)}$$

It can be shown that in cases where the correlation matrix $\underset{\sim}{C}(t)$ (see below) is symmetric, which holds e.g. in thermal equilibrium (c.f. Lax, 1960), the relation

$$\underset{\sim}{\sigma}\, \tilde{\underset{\sim}{M}} = \underset{\sim}{M}\, \underset{\sim}{\sigma} \qquad (B.3.23)$$

is valid. If det $\underset{\sim}{M} \neq 0$ (rank $\underset{\sim}{M} = n$) and hence $\underset{\sim}{M}^{-1}$ exists, the solution of the system of equations (B.3.20) may drastically be simplified to

$$\underset{\sim}{\sigma}^2 = 1/2\, \underset{\sim}{M}^{-1}\, \underset{\sim}{B}(\underset{\sim}{N}^s) \qquad (B.3.24)$$

Correlation matrix

In extension of the autocorrelation function, a correlation matrix $\underset{\sim}{C}(t)$ describing the auto- and crosscorrelations of the fluctuating parts α_i of the Markovian variables N_i is introduced.

$$\underset{\sim}{C}(t) = \langle \underset{\sim}{\alpha}(t)\, \tilde{\underset{\sim}{\alpha}}(0) \rangle = \sum_{\text{all } \underset{\sim}{\alpha}, \underset{\sim}{\alpha}'} \underset{\sim}{\alpha}(t)\, \tilde{\underset{\sim}{\alpha}}(0) P(\underset{\sim}{N}, t; N(0)) \qquad (B.3.25)$$

Often the correlation matrix is defined by $\langle \underset{\sim}{\alpha}(0)\, \tilde{\underset{\sim}{\alpha}}(t) \rangle$ and is just the transpose of $\underset{\sim}{C}(t)$ in (B.3.25).

After simple analytical manipulations of (B.3.25) $\underset{\sim}{C}(t)$ can be expressed by the variance matrix and the fundamental solutions $\underset{\sim}{\Omega}(t)$ or $\bar{\underset{\sim}{\Omega}}(t)$ of the linearized equations (B.3.13):

$$\underset{\sim}{C}(t) = \bar{\underset{\sim}{\Omega}}(t)\, \underset{\sim}{\sigma}^2 = \underset{\sim}{\Omega}(t)\, \underset{\sim}{\sigma}^2 \qquad (B.3.26)$$

The latter equality follows directly from the definition (B.3.19) of $\underset{\sim}{\sigma}^2$ and by comparison of (B.3.15) with (B.3.17).

Noise Spectrum Matrix

The noise spectrum matrix $G(\omega)$ is defined by (c.f.(B.1.7):

$$\underset{\sim}{G}(\omega) = 2 \text{ Re} \int_0^\infty (\underset{\sim}{C}(t) + \bar{\underset{\sim}{C}}(t))\, e^{i\omega t} dt \tag{B.3.27}$$

(i: imaginary unit). With the use of the representation (B.3.26) of $\underset{\sim}{C}(t)$ by the matrix exponential one gets from (B.3.27)

$$\underset{\sim}{G}(\omega) = 2 \text{ Re} \left[(\underset{\sim}{M} - i\omega\underset{\sim}{E})^{-1} \underset{\sim}{\sigma}^2 + \underset{\sim}{\sigma}^2 (\tilde{\underset{\sim}{M}} + i\omega\underset{\sim}{E})^{-1} \right] \tag{B.3.28}$$

or, as has been shown by Chen (1975):

$$\underset{\sim}{G}(\omega) = 2 \left[(\underset{\sim}{M}^2 + \omega^2\underset{\sim}{E})^{-1} \underset{\sim}{M}\underset{\sim}{\sigma}^2 + \underset{\sim}{\sigma}^2 (\tilde{\underset{\sim}{M}}^2 + \omega^2\underset{\sim}{E})^{-1} \tilde{\underset{\sim}{M}} \right] \tag{B.3.29}$$

(B.3.28) and (B.3.29) may be simplified for time reversible systems. Time reversibility means

$$\underset{\sim}{C}(t) = \underset{\sim}{C}(-t) \tag{B.3.30a}$$

or equivalently symmetry of $\underset{\sim}{C}$

$$C_{ik}(t) = C_{ki}(t) \tag{B.3.30b}$$

Furthermore the validity of (B.3.30) is equivalent with the validity of (B.3.23). Then, if $\underset{\sim}{M}^{-1}$ exists, $\underset{\sim}{\sigma}^2$ can be replaced according to (B.3.24) by $\underset{\sim}{M}$ and $\underset{\sim}{B}(\underset{\sim}{N}^s)$ and $\underset{\sim}{G}(\omega)$ becomes

$$\underset{\sim}{G}(\omega) = 2(\underset{\sim}{M}^2 + \omega^2\underset{\sim}{E})^{-1}\, \underset{\sim}{B}(\underset{\sim}{N}^s) \tag{B.3.31}$$

Naturally, $\underset{\sim}{G}(\omega)$ may directly be calculated from (B.3.27) with the use of the fundamental matrix $\underset{\sim}{\Omega}(t)$. If the eigenvalues of $\underset{\sim}{M}$ are all real (nonpositive), the components of $\underset{\sim}{\Omega}(t)$ and $\underset{\sim}{C}(t)$ consists of a sum of relaxation terms and the components of $\underset{\sim}{G}(\omega)$ are a sum of Lorentz spectra. If $\underset{\sim}{M}$ has also complex eigenvalues, $\underset{\sim}{\Omega}$ and $\underset{\sim}{C}$ may show a damped oscillatory behaviour and $\underset{\sim}{G}(\omega)$ may have peaks (c.f. Chen, 1975, Frehland and Stephan, 1978). Complex eigenvalues can only occur at non-equilibrium, e.g. for driven cycling reactions (see e.g. Higgins, 1967, Hearon, 1953, Chen, 1975).

3.5 Correlations and Spectra of Measurable Quantities

In many cases the variables N_i cannot be measured, but a set of observable (not necessarily Markovian) quantities X_β, $\beta = 1,2,\dots,r$, exists which is given by linear mapping of the state variables N_i:

$$X_\beta = \sum_{i=1}^{n} \alpha_{\beta i} N_i \qquad (B.3.32)$$

The measurable time correlations are determined by the time correlation matrix $\underset{\sim}{C}(t)$ (Chen, 1978):

$$C_{\beta\gamma}(t) := \overline{X_\beta(0)\, X_\gamma(t)} = \sum_{i,k=1}^{n} \alpha_{\beta i}\, C_{ik}(t)\, \alpha_{\gamma k} \qquad (B.3.33)$$

Correspondingly the spectral densities $S_{\beta\gamma}(\omega)$ are given by

$$S_{\beta\gamma}(\omega) = \sum_{i,k=1}^{n} \alpha_{\beta i}\, \alpha_{\gamma k}\, G_{ik}(\omega) \qquad (B.3.34)$$

3.6 Direct Calculation of Fluctuations from the Fundamental Solutions of the Master Equation

In many cases a tractable solution of the master equation in the form (B.3.7) is not available, though general results about e.g. the existence and stability of a time independent solution are known. Hence very often it is sensible to use the approach presented, which starts from the linearized phenomenological equations. Nevertheless, for reasons of principle, which will also be discussed below in part C in the systematic treatment of discrete transport systems, it is worthwhile to write down the fluctuation properties of the system with the use of the fundamental solution matrix $\underset{\sim}{\Omega}$ of the master equation. This leads to a remarkable simplification of the corresponding variance matrix and the correlation and noise spectrum matrices. Furthermore, these results seem to be useful e.g. in those cases, where the system consists of a number N_p of identical noninteracting subunits. Then the master equation for a single subunit can be treated and the fluctuations generated by the whole system are given by N_p times the fluctuations of one subunit. In the case of membrane channel noise, which is shortly discussed below, this can be done under the assumption that the channels act independently.

Starting from the master equation in the form (B.3.7), the variances, correlations and spectra can formally be introduced in the same way as in sections 3.4 and 3.5, where we started from the phenomenological equations (B.3.10). Hence all results for $\underset{\sim}{\sigma}^2$, $\underset{\sim}{C}$, and $\underset{\sim}{G}$ are valid also in this case. But the state variables N_ν now have a more 'microscopic' meaning and can take only the values 1 or 0 according as the system is in the corresponding state or is not. This fact leads to a simplification of the structures of $\underset{\sim}{\sigma}^2$, $\underset{\sim}{C}$ and $\underset{\sim}{G}$.

First we determine the correlations $\langle N_\mu N_\nu \rangle$ with the use of the definitions of ensemble averages introduced in sections 2.1 and 2.2 of part A. Taking into account, that N_μ can assume the values 0 and 1 only, we get simply

$$\langle N_\mu N_\mu \rangle = P_\mu^s \qquad \text{(B.3.35a)}$$

with P_μ^s as the stationary solution of the master equation (B.3.7). For determination of the crosscorrelations $\langle N_\mu N_\nu \rangle$, we use that for $\mu \neq \nu$ the joint probability $P(N_\mu = 1; N_\nu = 1)$ vanishes, because the system cannot be in two different states at one time.
Hence

$$\langle N_\mu N_\nu \rangle = 0 \quad \text{for} \quad \mu \neq \nu \qquad \text{(B.3.35b)}$$

With the definition (A.1.5) of $\underset{\sim}{\sigma}^2$, from (B.3.35) it follows that

$$\sigma_{\mu\nu}^2 = \delta_{\mu\nu} P_\mu^s - P_\mu^s P_\nu^s \qquad \text{(B.3.36)}$$

Because the state variables N_μ assume only the values 0 and 1, the diagonal components of $\underset{\sim}{\sigma}^2$ can be derived from the results in section A.3 for the Binomial distribution.

With (B.3.36) the correlation matrix in (B.3.26) can be simplified to

$$C_{\mu\nu}(t) = \Omega_{\mu\nu}(t) P_\nu^s \qquad \text{(B.3.37)}$$

where we have used that according to (B.3.18) the fundamental solution matrix does not contain the stationary solutions and hence

$$\underset{\sim}{\Omega} \, \underset{\sim}{\tilde{P}}^s = 0 \qquad \text{(B.3.38)}$$

The noise spectrum matrix follows from (B.3.27) and (B.3.37)

$$G_{\mu\nu}(\omega) = 2 \int_0^\infty (\Omega_{\mu\nu}(t)P^s_\nu + \Omega_{\nu\mu}(t)P^s_\mu) \cos \omega t \, dt \qquad (B.3.39)$$

or from (B.3.29) with the explicit form (B.3.36) of $\underset{\sim}{\sigma}^2$.

The correlations and spectra of measurable quantities are derived as shown in section 3.5. In case the system consists of a number N_p of identical noninteracting subunits and $\underset{\sim}{\Omega}$ is the fundamental solution matrix of the master equation for one single subunit, (B.3.35), (B.3.36), (B.3.37) and (B.3.39) have to be multiplied by the factor N_p.

3.7 Current Noise Generated by the Opening and Closing of Channels

In the last years a number of papers and reviews has been published on the theoretical treatment of electrical noise generated by the opening and closing of ionic channels in biological membranes (see e.g. Verveen a. Derksen (1965), Verveen a. de Felice (1974), Conti a. Wanke (1975), Neher a. Stevens (1977), de Felice (1977), Chen (1978), De Felice (1981), Conti a. Carbone (1981), Kolb (1981)).
We will demonstrate at the simple example of channels with one conducting state, how the presented master equation approach can be applied. The ionic transport system in the membrane is assumed to consist of a number N_p of identical transport subunits (channels) which are permeable only for one ion species and act independently.

The basic assumption is that the measured current is proportional to the number of channels in the conducting 'open' state. The assumption seems to be sensible in those cases where the transport of ions through the channels is fast compared with the open-closed kinetics of the channels. In part C, from a more general treatment of current noise, we will show that under this condition the assumption can be justified. In nerve membrane channels the transport kinetics obviously are some orders of magnitude faster than the channel open-closed kinetics.

It is further assumed that the single channel can be in a number of different states. The channel kinetics are determined by the master equation for the single channel. The open state is denoted by the index 1. Hence the measured current J is given by

$$J = N_p\, j^s\, P_1(t) = j^s\, N_1(t) \quad , \qquad (B.3.40)$$

$P_1(t)$: Solution of the master equation, probability of the channel to be in the open state; j^s: current through a single open channel, which for a given voltage can be assumed to be stationary. As stated above, the current fluctuations are given by the current fluctuation properties of a single channel times N_p. And according to the basic assumption the current fluctuations primarily are given by the number of fluctuations in the number N_1 of open channels.

Hence, applying (B.3.32) where we have to set n=1 and $\alpha_{11}=j^s$, and using the results of section 3.6 we get the variance σ_J^2, autocorrelation function $C_{\Delta J}(t)$ and spectral density $G_{\Delta J}(\omega)$ of current fluctuations:

$$\sigma_J^2 = {j^s}^2 \; N_p \; P_1^s(1-P_1^s)$$

$$C_{\Delta J}(t) = {j^s}^2 \; N_p \; P_1^s \; \Omega_{11}(t)$$

$$G_{\Delta J}(\omega) = {j^s}^2 \; N_p \; P_1^s \; 4 \int_0^\infty \Omega_{11}(t) \cos \omega t \, dt \qquad (B.3.41)$$

with the component Ω_{11} of the fundamental solution matrix for the single channel master equation, denoting the deviation of the probability $P_1(t)$ that the channel is open under the condition that at time t=0 it was open, from the stationary probability $P_1^s = \lim_{t\to\infty} P_1(t)$.

Up to now we have not yet specified the special structure of the master equation. In case the nonzero eigenvalues are real, the auto-correlation function has a pure relaxation and the noise spectrum a pure Lorentzian structure.

For the simple two state model with state 2 as the closed state the explicit structure of the master equation is

$$\frac{d\,P_1}{dt} = -k_1 P_1 + k_2 P_2 = -\frac{d\,P_2}{dt} \qquad \text{(B.3.42)}$$

with the solution

$$P_1^s = \frac{k_2}{k_1+k_2}$$

$$\Omega_{11}(t) = e^{-\frac{t}{\tau}}.\tau k_1\,, \qquad \tau = \frac{1}{k_1+k_2} \qquad \text{(B.3.43)}$$

Then from (B.3.42) follows the well known result for the current fluctuations generated by the channel open-closed kinetics

$$\sigma_J^2 = N_P \frac{k_1\,k_2}{(k_1+k_2)^2}\,j^{s\,2}$$

$$C_{\Delta J}(t) = N_P\, j^{s\,2} \frac{k_1\,k_2}{(k_1+k_2)^2}.e^{-\frac{t}{\tau}}$$

$$G_{\Delta J}(\omega) = 4\,N_P\, j^{s\,2} \frac{k_1 k_2}{(k_1+k_2)^2}.\left(\frac{\tau}{1+\omega^2\tau^2}\right) \qquad \text{(B.3.44)}$$

C. Transport Fluctuations Around Steady States

1. Introductory Remarks on Transport Fluctuations

1.1. Characterization

In part B we have described the theoretical approach to fluctuations of (Markovian) variables around stable steady states. These approaches are usually applied to the investigation of nerve channel noise with the use of the basic assumption that the electric current at any instant is proportional to the number of open channels for two state models and correspondingly for channels with different conducting states. This assumption is plausible in cases where the movement of ions through the channels is fast compared with the typical times of the channel open-closed kinetics.

Nevertheless it should be kept in mind that primarily the electric current is always generated by the movement or transport of charges. In the barrier channel model discussed below the movement of charges is given by the jumps of the ions over energy barriers separating different binding sites.

From now on, the electric fluctuations, that are generated by movement or displacement of charges, will be called transport noise. Primarily, in this definition all electric noise is transport noise. In part C we will present a theoretical approach to transport fluctuations which generally may be applied to the treatment of electric noise in biological systems. The aim of part C is twofold: First, we shall try to prepare special applications to noise experiments by the explicit treatment of a number of different transport systems. Second, we shall use the formalism for the investigation of general properties of stochastic transport fluctuations at non-equilibrium.

The latter aspect will be subject of section 7, where we discuss why at non-equilibrium the fluctuation dissipation theorem (Nyquist theorem), relating the fluctuations to the macroscopic response properties (admittance), breaks down. An apparent discrepancy between our result and a number of classical papers, where generalizations of the fluctuation dissipation theorem to non-equilibrium steady states are derived, will be clarified: while these papers are concerned with fluctuations of scalar fluxes, i.e. the time derivatives of state variables, our treatment of directed transport is based on vectorial fluxes. Therefore as a main result we will show the essentially different behaviour of scalar and vectorial quantities at nonequilibrium.

As in part B we will mainly restrict to the description of transport in discrete systems. As we will show at a number of special examples, the discrete description seems to be adequate for systems with discontinuous structures and coupling between transport and other processes as e.g. chemical reactions. It has been successful for the theoretical description of a number of complex models of ion transport through biological membranes. Furthermore the discrete description is more general than a continuum description in so far as continuum models are derivable from discrete models as limiting cases.

In section 2 we will present the general concept of transport in discrete systems. In the most general case we start from the master equation (B.3.7). For the treatment of systems with negligible interactions in section 5 it is a great simplification to use the linearized phenomenological equations.

1.2 A Simple Example: Channels with One Binding Site

We discuss the special case of permanently open pores (channels) with one binding site within the pore, one ion species and negligible interactions. At this simple example we give a first insight into the typical problems and difficulties concerning the theoretical treatment of transport fluctuations. The transport system is considered to consist of a great number of identical noninteracting channels, which are built into a membrane separating two aqueous solutions. In Fig. CO the potential profile within one

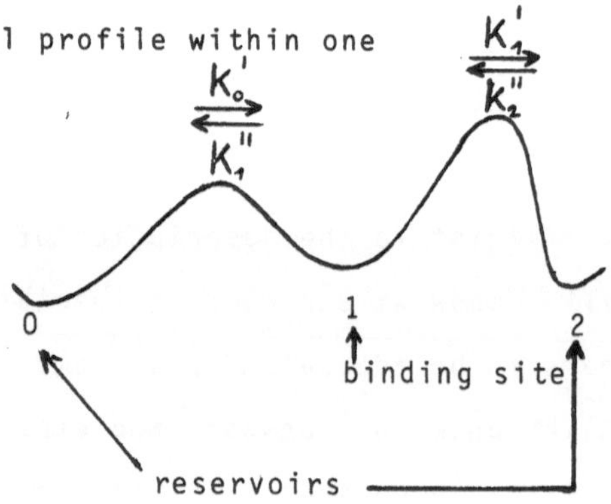

Fig. CO Potential profile in a channel with one binding site.

channel is indicated. The binding site within the channel (denoted by index 1) is separated by energy barriers from the reservoirs on the left- and right-hand sides (denoted by indices 0 and 2). The variable N_1 denotes the occupation number of ions at the binding site. If the number of unoccupied channels at any instant is great compared with the number of channels occupied by an ion, interactions between the ions may be neglected and the phenomenological equations (B.3.10) reduce to one equation describing the time dependence of $\langle N_1 \rangle$:

$$\frac{d \langle N_1 \rangle}{dt} = k_0' N_0 - (k_1' + k_1'') \langle N_1 \rangle + k_2'' N_2 \qquad (C.1.1)$$

N_o, N_2 are the constant concentrations (reservoirs) at the left- and right-hand sides respectively, k_o', k_2'' the corresponding rate constants for jumps of the ions into the pores and k_1', k_1'' for jumps out of the pores to the right, left respectively. The fundamental solution Ω_{11} of the homogenous equation belonging to (C.1.1) is simply

$$\Omega_{11} = e^{-t/\tau} \quad , \qquad \tau = (k_1' + k_1'')^{-1} \quad , \tag{C.1.2}$$

and the stationary solution

$$N_1^s = (k_o' N_o + k_2'' N_2) \cdot \tau \tag{C.1.3}$$

The four individual fluxes ϕ_{01}, ϕ_{10}, ϕ_{12}, ϕ_{21} denote the jumps of ions from the binding site to the left reservoir, from the left and right reservoirs into the pore, and from the binding site to the right reservoir respectively. We introduce the notations

$$\phi_{10} = \phi_1', \ \phi_{01} = \phi_1''$$
$$\phi_{21} = \phi_2', \ \phi_{12} = \phi_2'' \tag{C.1.4}$$

Furthermore we define the total fluxes ϕ_1, ϕ_2 over the first and second barrier

$$\phi_1 = \phi_1' - \phi_1'' \quad , \qquad \phi_2 = \phi_2' - \phi_2'' \tag{C.1.5}$$

In order to find an expression for the measured electric current J, we imagine that ionic jumps over the first and second barrier yield well defined contributions to J. Hence we can write

$$J = \gamma_1 \phi_1 + \gamma_2 \phi_2 \qquad \text{a)}$$
$$\gamma_1 + \gamma_2 = ze_o \qquad \text{b)} \tag{C.1.6}$$

(C.1.6) holds, because transport of an ion over the two barriers means the transport of a charge ze_0 across the membrane. Because interactions are neglected, the expectation values of the fluxes are given by the rate constants times the number of ions at the corresponding binding site:

$$<\phi_1'> = k_0' N_0 = \phi_1'^s \quad , \qquad <\phi_1''> = k_1'' <N_1>$$
$$<\phi_2'> = k_1' <N_1> \quad , \qquad <\phi_2''> = k_2'' N_2 = \phi_2''^s \tag{C.1.7}$$

Because the finite duration of an ionic jump is neglected, the 'microscopic' fluces consist of a sum of δ-pulses, each of which is generated by an ionic jump.

While the expectation values and the macroscopic fluxes are determined by $<N_1>$, the microscopic fluxes are not uniquely determined by N_1. Though there is a dependence on the time derivative $\dot{N}_1$, a change in N_1 can be caused by a jump of an ion over the first <u>or</u> the second barrier. Only in special cases (see the discussion of this point in section 2) a unique relation between the time derivatives of the occupation numbers and the current exists and the treatment of transport fluctuations can be reduced to the treatment of the fluctuations in the occupation numbers. In general we need another theoretical approach to transport fluctuations.

For this special example of noninteracting pores with one binding site Läuger (1975) has presented a combinatorial approach by distinction of four different possible classes of pairs of pulses which consist of a jump of an ion into the pore (from the left or right) and a jump out of the pore (to the left or to the right). By application of Carson's theorem (B.2.9), (B.2.10) to these four mutually independent classes of events he derived an expression for the spectral density $G_{\Delta J}(\omega)$ of fluctuations in the fluctuating part

$$\Delta J = J - J^{s} \tag{C.1.8}$$

of current J (J^{s}: stationary current), which in a slightly modified form reads (c.f. Frehland, 1978)

$$G_{\Delta J}(\omega) = 2A_1 + A_2 \frac{4\tau}{1+\omega^2\tau^2} \tag{C.1.9}$$

with

$$A_1 = \gamma_1^2(k_o'N_o + k_1''N_1^s) + \gamma_2^2(k_1'N_1^s + k_2''N_2)$$
$$A_2 = (\gamma_1 k_o'N_o - \gamma_2 k_2''N_2)\,(\gamma_2 k_1' - \gamma_1 k_1'') \tag{C.1.10}$$

The corresponding autocorrelation function $C_{\Delta J}(t)$ can according to the Wiener-Khintchine relations be calculated from $G_{\Delta J}(\omega)$ by Fourier-transformation:

$$C_{\Delta J}(t) = A_1 \cdot\delta(t) + A_2\,\Omega_{11}(t)\,,$$

$$\Omega_{11} = e^{-t/\tau} \tag{C.1.11}$$

The fluctuations for this simple example show properties which are typical of transport noise in discrete systems. The spectrum (autocorrelation function) consists of a frequency independent (delta-shaped) term, directly generated by the jumps of the discrete charges between discrete states, and a frequency-dependent (time-dependent) term, which is determined by the (fundamental) solution of the phenomenological equations and comes from the (averaged) behaviour of the system as consequence of jumps of the ions. The first term yields the high frequency white noise limit. The second term is called Lorentzian noise or $1/f^2$

noise because it is proportional to the reciprocal square of frequency at frequencies higher than the characteristic 'frequency' $1/2\pi\tau$. For low frequencies ($\omega \to o$) the spectrum is white as well. As we shall discuss below, the low frequency limit of transport noise at equilibrium is always and at non-equilibrium is often lower (or equal) than the high frequency limit as consequence of the linear response properties of the systems.

For very low frequencies ($\omega \to o$) and high applied voltage (i.e. k_1'', $k_2'' \to o$) we get from (C.1.6), (C.1.9), (C.1.10) and using that the stationary unidirectional fluxes $k_o'N_o$ and $k_1'N^s$ are equal:

$$G_{\Delta J}(\omega \to o) = k_o'N_o$$

$$= 2\ ze_o\ J^s \qquad (C.1.12)$$

As expected the result (C.1.13) agrees with Schottky's shot noise theorem (B.2.21), because the low frequency limit is determined by the long-time behavior. And for long times correlations between the ionic jumps disappear.

2. The Concept of Transport in Discrete Systems

Generally, we start from a master equation (B.3.7) for the whole system or for a subunit of the system in case the system consists of a number of identical noninteracting transport subunits. If interactions can be neglected the linearized phenomenological equations can be used as basic equations.

2.1 The Fluxes

The master equation in the form (B.3.7) reads

$$\frac{\partial P_\mu}{\partial t} = \sum_{\nu,\mu\neq\nu} (M_{\mu\nu}P_\nu - M_{\nu\mu}P_\mu) \qquad (C.2.1)$$

$M_{\mu\nu}$ $(\mu\neq\nu)$: transition rate per unit time

According to the introductory remarks in the preceding section transport must be coupled to changes in the system, i.e. to transitions between different states. We introduce the unidirectional 'fluxes' $\phi_{\mu\nu}$ between different states μ and ν $(\nu\neq\mu)$. These fluxes are built up by the transitions $\nu\rightarrow\mu$ the duration of which is assumed to be very short and therefore negligible. If at times $t_i(\mu,\nu)$ occur transitions $\nu\rightarrow\mu$, then $\phi_{\mu\nu}$ is given by a sum of δ-pulses.

$$\phi_{\mu\nu} = \sum_i \delta(t - t_i(\mu,\nu)) \qquad (C.2.2)$$

The matrix $\underset{\sim}{\phi}$ with components $\phi_{\mu\nu}$ is called the flux matrix $(\phi_{\nu\nu} = 0)$.

Because a δ-pulse at $t = t_i(\mu,\nu)$ corresponds to an increase of the state variable N_μ by one from zero to one and decrease of N_ν by one from one to zero, the time derivatives of N_ν are uniquely given by the fluxes through the balance equation

$$\dot{N}_\nu = \sum_\mu (\phi_{\nu\mu} - \phi_{\mu\nu}) \qquad (C.2.3)$$

Graphtheoretical methods can be useful for the analysis of master equation systems (Schnakenberg, 1976). The nodes of the graph are represented by the states N_ν and the edges by the transition rates $M_{\mu\nu}$ or by the fluxes $\phi_{\mu\nu}$ $(\mu \neq \nu)$. $\underset{\sim}{N}$ is regarded as a scalar field in the network and $\underset{\sim}{M}$, $\underset{\sim}{\phi}$ as vector fields. Analogously as for continuous fields the operators gradient, rotation and divergence can be defined. E.g. the divergence of $\underset{\sim}{\phi}$ at state ν is defined as

$$\text{div } \underset{\sim}{\phi}(\nu) = \sum_\mu (\phi_{\nu\mu} - \phi_{\mu\nu}) \qquad (C.2.4)$$

Comparison with (C.2.3) yields

$$\text{div } \underset{\sim}{\phi}(\nu) = \dot{N}_\nu \qquad (C.2.5)$$

(C.2.5) shows the <u>vectorial</u> character of the introduced fluxes. Though the time derivatives of N_ν are according to (C.2.3), (C.2.5) uniquely determined by the fluxes, a unique representation of fluxes or related quantities by the $\dot{N}_\nu$ is in general not possible. This problem will be further discussed below in section 2.3.

Because of the probabilistic nature of laws which determine the transitions, equations describing the time-dependent behaviour of fluxes are available only for the averaged or expected fluxes $<\phi_{\mu\nu}>$:

$$<\phi_{\mu\nu}> = M_{\mu\nu} P_\nu \tag{C.2.6}$$

Nevertheless the 'microscopic' picture of the fluxes as being composed of δ-pulses must be kept in mind and it is decisive for a satisfactory treatment of the stochastic behaviour (fluctuations) of transport phenomena.

Equilibrium - Nonequilibrium

With the use of the expected stationary fluxes

$$\phi_{\mu\nu}^s = M_{\mu\nu} P_\nu^s \tag{C.2.7}$$

we define the system to be in an equilibrium steady state if the stationary flux matrix $\tilde{\phi}^s$ is symmetric, i.e. satisfies detailed balance

$$\phi_{\mu\nu}^s = \phi_{\nu\mu}^s \quad \text{for all } \mu,\nu \tag{C.2.8}$$

(C.2.8) means that on the average the number of transitions $\mu \rightarrow \nu$ equals the number of transitions $\nu \rightarrow \mu$. If $\tilde{\phi}^s$ is not symmetric the stationary state is called a nonequilibrium steady state.

As we shall see below the properties of transport fluctuations at equilibrium and nonequilibrium are essentially different. From the theory of chemical reactions (e.g. Bak (1959), Higgins (1967)) it is well-known that the validity of detailed balance means the validity of the Wegscheider conditions for $M_{\mu\nu}$.

2.2 Current as a Transport Observable

Up to now we have introduced the state variables N_ν, probabilities P_ν, fluxes $\phi_{\mu\nu}$ and transition rates $M_{\mu\nu}$. In most cases the variables N_ν or $\phi_{\mu\nu}$ are not measurable. This will be shown explicitely with the examples for ionic transport systems in membranes. Hence we now develop the concept of transport observables, which makes possible a satisfactory formal treatment of measurable transport processes in discrete systems. In the following, for simplicity, we exclude cases of multiple connections between different states. But these cases can be treated in complete analogy (Frehland,1981, and the example of channels with one binding site treated in section 6.3 of part C).

We define a transport observable T as a linear mapping of $\underset{\sim}{\Phi}$ mediated by a matrix $\underset{\sim}{\gamma}$ with vanishing diagonal elements

$$T = \sum_{\substack{\mu,\nu \\ \mu \neq \nu}} \gamma_{\mu\nu} \phi_{\mu\nu} \tag{C.2.9}$$

Transport is determined by transitions in the system and a special transition $\nu \rightarrow \mu$ is coupled with a special, well defined contribution to the transport observable T. All examples below shall be restricted to transport of electric charges, where T is the electric current J measured by a suitable experimental set-up though also other transport processes, e.g. transport of mass, can be treated in the same way (c.f. Frehland, 1981). Then obviously $\underset{\sim}{\gamma}$ is antisymmetric

$$\gamma_{\mu\nu} = -\gamma_{\nu\mu} \tag{C.2.10}$$

(C.2.10) expresses a balance relation, which leads to a strong coupling between $\underset{\sim}{\phi}$ and J. E.g. at equilibrium J^S vanishes.

2.3 Conservative Systems, Open Systems

A subclass of transport systems is defined by the condition on $\underset{\sim}{\gamma}$ that for arbitrary states ν, μ and an arbitrary sequence of states leading from μ to ν the sum over all $\gamma_{\alpha\beta}$ belonging to the sequence vanishes. If $\underset{\sim}{\gamma}$ satisfies this condition it is called (in analogy to continuum systems) to have vanishing rotation

$$\text{rot}\ \underset{\sim}{\gamma} = 0 \tag{C. 2.11}$$

In this case, as in continuum systems, $\underset{\sim}{\gamma}$ may be derived from a scalar 'potential'

$$u(\nu) := u_\nu \tag{C.2.12a}$$

through

$$\gamma_{\mu\nu} = u_\mu - u_\nu \tag{C.2.12b}$$

We call systems satisfying (C.2.11) conservative with respect to $T(\gamma)$. (C.2.11) with (C.2.12) includes antisymmetry (C.2.10) of $\underset{\sim}{\gamma}$. For conservative transport the following theorem holds:

Theorem: If and only if (C.2.11) or (C.2.12) are valid, the transport observable $T(\underset{\sim}{\gamma})$ may be represented as the time derivative of a linear combination of the state variables N_ν

$$T = \frac{dU}{dt}, \qquad U = \sum_\nu u_\nu N_\nu \tag{C.2.13}$$

Proof: Consider an arbitrary transition $\beta \to \alpha$, occurring at t'. As consequence of this transition the value of N_α increases by one, the value of N_β decreases by one and all other state variables remain unchanged. Hence for a sufficiently small time interval Δt around t', where only one transition $\beta \to \alpha$ occurs, the change of U in (C.2.13) is simply

$$U(t_2 > t', t_2 - t' < \Delta t) - U(t_1 < t', t' - t_1 < \Delta t) = u_\alpha - u_\beta = \gamma_{\alpha\beta}$$

Therefore

$$\frac{dU}{dt}(t, |t - t'| < \Delta t) = \gamma_{\alpha\beta} \cdot \delta(t-t')$$

which is just the δ-shaped contribution of a transition $\beta \to \alpha$ to the transport observable T. Q.e.d.

(C.2.13) holds for the expected values of T as well:

$$\langle T \rangle = \sum_{\mu\nu} \gamma_{\mu\nu} \langle \phi_{\mu\nu} \rangle = \sum_{\nu} u_\nu \langle \dot{N}_\nu \rangle \qquad \text{(C.2.14)}$$

This means that the stationary expected value $\langle T^s \rangle$ of the transport variables T vanishes:

$$\langle T^s \rangle = \sum_{\nu} u_\nu \langle \dot{N}_\nu^s \rangle = 0 \qquad \text{(C.2.15)}$$

The rotation of $\underset{\sim}{\gamma}$ is nonvanishing if and only if there exist at least two states μ, ν and two sequences $C_1^{\mu\nu}$ and $C_2^{\mu\nu}$ of transitions leading from μ to ν where the sums over the corresponding $\gamma_{\alpha\beta}$ are different. We call a system <u>open with respect to $T(\gamma)$</u>, if it is in contact with the environment (reservoirs) in so far as an exchange of the transported stuff represented by T with outer reservoirs is possible. For antisymmetric $\underset{\sim}{\gamma}$ this definition leads to the following sufficient condition:

> A discrete transport system is open with respect to T if rot $\underset{\sim}{\gamma} \neq 0$ and $\underset{\sim}{\gamma}$ is antisymmetric.

The following arguments may show that this criterion is reasonable The sum over $\gamma_{\alpha\beta}$ belonging to the closed path formed by $C_1^{\mu\nu}$ and $C_2^{\mu\nu}$

is nonzero $\sum_{C_1^{\mu\nu}} \gamma_{\alpha\beta} - \sum_{C_2^{\mu\nu}} \gamma_{\alpha\beta} \neq 0$

As consequence one gets a nonzero contribution to T, if the system runs through this closed path, though before and afterwards the system is in the same state. This is interpreted as a taking up (or delivery) of the transported stuff represented by T from (or into) outer reservoirs. Furthermore these arguments lead to the proposition of the same condition to be a necessary criterion for transport systems to allow nonvanishing steady state transport $T^s \neq 0$ through the system. Obviously a further necessary condition is that the steady state of the system is a nonequilibrium state.

2.4 Interactions Neglected

In case interactions are neglected, the use of the linearized phenomenological equations yields a remarkable simplification because the number of variables can drastically be reduced. Instead of state variables now the system can be characterized by a set of n occupation numbers N_i (i = 1,2,...,n) of ions at different places. We now use latin indices for numbering these different places.
The linear phenomenological equations (B.3.10) are

$$\frac{d < \underset{\sim}{N}(t) >_{\underset{\sim}{N}(0)}}{dt} = \underset{\sim}{M} < \underset{\sim}{N}(t) >_{\underset{\sim}{N}(0)} + \underset{\sim}{Y} \quad (C.2.16)$$

with the stationary solution $\underset{\sim}{N}^s$ of

$$\underset{\sim}{M}\,\underset{\sim}{N}^s + \underset{\sim}{Y} = 0 \quad (C.2.17)$$

Eqns. (C.2.16) and (C.2.17) are assumed to describe also the macroscopic behaviour of the system, e.g. the time dependent relaxation of $\underset{\sim}{N}(t)$ after a macroscopic disturbance ($\underset{\sim}{N}(0) - \underset{\sim}{N}^s$) of the system (regression hypothesis). The general solution with the use of the fundamental matrix $\underset{\sim}{\Omega}(t)$ has been described in section B.3.

Analogously as above we can introduce the unidirectional 'fluxes' ϕ_{ik}, i.e. number of jumps $k \rightarrow i$. They build up the flux matrix $\underset{\sim}{\phi}$. ϕ_{ii} is set equal to zero. Because the duration of (ionic) jumps is neglected, ϕ_{ik} consists of a sum of delta-shaped pulses, each of which is generated by a transition $k \rightarrow i$. The expectation and macroscopic values of fluxes are assumed to be given by $\langle \underset{\sim}{N}(t) \rangle$ and the matrix $\underset{\sim}{M}$ of coefficients through

$$\langle \phi_{ik}(t) \rangle = M_{ik} \langle N_k(t) \rangle, \qquad i \neq k \qquad (C.2.18)$$

and the stationary fluxes ϕ_{ik}^s forming the stationary flux matrix $\underset{\sim}{\phi}^s$

$$\phi_{ik}^s = M_{ik} N_k^s \qquad (C.2.19)$$

Because interactions are neglected the validity of (C.2.18) and (C.2.19) is obvious: the probability for transitions $k \rightarrow i$ is proportional to the occupation number N_k and M_{ik} is the jump rate from k to i.

If $\underset{\sim}{Y} \neq 0$, the system is in contact with (m-n) outer reservoirs and it is necessary to take into account also fluxes into, from and between these reservoirs by extending the (nxn)-matrix to a (mxm)-matrix $\underset{\sim}{\overline{\phi}}$ (m>n) with the components $\phi_{\mu\nu}$ and introducing the rate constants $M_{\mu i}$, $M_{i\mu}$, $M_{\mu\nu}$ (μ,ν>n) for the corresponding transitions E.g.

$$\langle \phi_{\mu i} \rangle = M_{\mu i} \langle N_i \rangle, \qquad \mu > n, \quad i \geq n \qquad (C.2.20)$$

is the expectation value of flux from i to the (μ-n)-th reservoir and $\phi_{\mu\nu}^{s}$ is the (stationary) flux between two reservoirs.

As above we arrive at an expression for the measured electric current: During the measurement the transport system is helt under constant voltage (voltage clamp). A special jump in the transport system is connected with a special contribution to the measured electric current J. Hence in correspondence to (C.2.9) J is given by the linear combination

$$J = \sum_{\substack{\mu,\nu=1 \\ \mu\neq\nu}}^{m} \gamma_{\mu\nu} \phi_{\mu\nu} \qquad \gamma_{\mu\nu} = -\gamma_{\nu\mu} \qquad \text{(C.2.21)}$$

For transitions $\nu\rightarrow\mu$, which yield no contribution to J, $\gamma_{\mu\nu}$ is set equal to zero.

The occurrence of an inhomogeneity $\underset{\sim}{Y} \neq 0$, taking into account the contact with reservoirs will yield a slight modification in the treatment of the fluctuations.

It is useful to extend the fundamental (n x n)-matrix $\underset{\sim}{\Omega}$ of the linear equations with components Ω_{ik} to a m x m-matrix with the components

$$\Omega_{\mu\nu} = \Omega_{ik} \qquad \text{for } \mu,\nu = i,k, \quad i,k \leq n$$

$$\Omega_{\mu\nu} = 0 \qquad \text{for } \mu \text{ or } \nu > n \qquad \text{(C.2.21)}$$

3. Transport Fluctuations at Equilibrium

As we shall show in section 4 the concept of transport in discrete systems allows a formally equal treatment of transport fluctuations around equilibrium and nonequilibrium steady states, though the properties of fluctuations at equilibrium will turn out to be essentially different from nonequilibrium situations. Before we present our approach in section 4, we discuss separately the equilibrium case.

3.1 The Nyquist or Fluctuation-Dissipation Theorem

Transport fluctuations at equilibrium can be treated with the use of the fluctuation-dissipation theorem (Callen, Welton (1951), Callen, Greene (1952), Kubo (1966)), which in the special case of electric fluctuations is called Nyquist theorem (Nyquist, 1928). By the Nyquist or fluctuation-dissipation theorem the microscopic fluctuations are expressed in terms of macroscopic response properties (dissipation, admittance) of the systems. E.g. from the frequency dependent complex admittance of a system the current or voltage fluctuations at equilibrium may directly be calculated:

$$G_{\Delta J}(\omega) = 4k_b T_a \, \mathrm{Re}\, Y(\omega) \qquad (C.3.1)$$

$$G_{\Delta V}(\omega) = 4k_b T_a \, \mathrm{Re}\left(\frac{1}{Y(\omega)}\right) \qquad (C.3.2)$$

k_b = Boltzmann constant; T_a = absolute temperature; Re $Y(\omega)$ = real part of the complex admittance $Y(\omega)$ of the system. The complex admittance is defined as follows: assume that additionally to the equilibrium potential a small (complex) periodic voltage

$$\varepsilon(t) = \varepsilon_0 \, e^{i\omega t} \qquad (C.3.3)$$

is applied to the system. The voltage $\varepsilon(t)$ is assumed to be small so that the corresponding frequency dependent current response $j(t,\omega)$ is linear in ε_0. Then $j(t,\omega)$ is given by

$$j(t,\omega) = \varepsilon_0 \, Y(\omega) \, e^{i\omega t} \quad , \qquad (C.3.4)$$

which defines the complex admittance $Y(\omega)$.
With the voltage applied to the system we have introduced an external force acting on the electric current via the voltage dependence of the rate constants.

The Nyquist theorem has been applied (Läuger (1978), Kolb a. Läuger (1979)) to electric equilibrium transport noise generated by open channels and carrier mediated ion transport, respectively. We will derive a general expression for the admittance in discrete transport systems.

3.2 Voltage Dependence of the Rate Constants

For the voltage dependence of the rate constants $M_{\mu\nu}$ we use the ansatz (Zwolinsky, Eyring a. Reese, 1948)

$$M_{\mu\nu}(V) = M_{\mu\nu}(V_0) \, e^{\alpha_{\mu\nu} \frac{u}{2}}$$

$$u = \frac{z \, e_0 (V - V_0)}{k_b T_a} \quad , \quad \alpha_{\mu\nu} = - \alpha_{\mu\nu} . \qquad (C.3.5)$$

ze_0 is the charge transported during a transition $\nu \rightarrow \mu$ in the transport system. The dimensionless number $\alpha_{\mu\nu}$ denotes the fraction $\Delta V_{\mu\nu}/(V-V_0)$ of the total applied voltage $(V-V_0)$ which is seen by $z \, e_0$ between states ν and μ. For small applied voltages $\underset{\sim}{M}$ may be linearly expanded:

$$M_{\mu\nu}(V_o + \varepsilon) = M_{\mu\nu}(V_o)\left[1 + \frac{\varepsilon}{2}\alpha_{\mu\nu}\frac{z\,e_o}{k_b T_a}\right] \tag{C.3.6}$$

If V_o is the equilibrium voltage, we write

$$\overline{M_{\mu\nu}} = M_{\mu\nu}(V_o) \tag{C.3.7}$$

For energetic reasons it is plausible to assume that the numbers $\alpha_{\mu\nu}$ are directly related to the constants $\gamma_{\mu\nu}$, which according to (C.2.9) or (C.2.21) determine the contribution of a transition $\nu \to \mu$ to the measured current J

$$\alpha_{\mu\nu} \cdot z\,e_o = \gamma_{\mu\nu} \tag{C.3.8}$$

Hence we get

$$M_{\mu\nu}(V_o + \varepsilon) = M_{\mu\nu}(V_o)\left[1 + \frac{\varepsilon}{2}\gamma_{\mu\nu}\frac{1}{k_b T_a}\right] \tag{C.3.9}$$

A heuristic formulation of relation (C.3.8) may be given by the following arguments (Frehland, 1980): When inside a transport system, which is helt under constant voltage V (voltage clamp), a charge q is transported between states μ and ν with a certain potential difference $\Delta_{\mu\nu}V$, then the corresponding loss (gain) of electric potential energy causes a change of the total voltage applied to the system. Because the system is to be helt under constant voltage, a charge $\Delta_{\mu\nu}q$ must be delivered from outside in order to compensate the change in voltage, generating a current pulse in the outer circuit. For energetic reasons the fraction $\frac{\Delta_{\mu\nu}q}{q}$ equals the fraction $\frac{\Delta_{\mu\nu}V}{V}$ and hence (C.3.8) is valid.

We wish to emphasize that the linear expansion (C.3.6) for the voltage dependence of the rate constants and the relation (C.3.8) between $\gamma_{\mu\nu}$ and $\alpha_{\mu\nu}$ are essential and explicitely used in the following general derivation of the admittance.

3.3 Linear Response and Complex Admittance

It is well known that (see e.g. Kubo, 1965) the linear response of a system to a small delta-shaped pulse of the external force is connected with the admittance analogously as spectral density and autocorrelation function by Fourier transformation: For a small voltage pulse

$$(V-V_0) = \zeta \cdot \delta(t) \qquad (C.3.10)$$

the resulting current response is j(t). Because of

$$\delta(t) = \int_{-\infty}^{+\infty} e^{i\omega t}\, d\omega \qquad (C.3.11)$$

and the linear properties of the system, the admittance $Y(\omega)$ is given by j(t) through

$$Y(\omega) = \frac{1}{\zeta} \int_{-\infty}^{+\infty} j(t) e^{-i\omega t}\, dt \qquad (C.3.12)$$

An expression for the linear response j(t) in discrete systems is derived as follows: According to (C.3.6) the voltage pulse (C.3.10) at t=0 effects a delta-shaped change

$$(M_{\mu\nu} - \overline{M_{\mu\nu}}) = \overline{M_{\mu\nu}}\, \frac{\zeta}{2}\, \frac{\gamma_{\mu\nu}}{k_b T_a}\, \delta(t) \qquad (C.3.13)$$

of the rate constant $M_{\mu\nu}$, which on the other hand according to (C.2.18) is connected with a small pulse in flux $\langle\phi_{\mu\nu}\rangle$

$$\frac{1}{2}\zeta\frac{\gamma_{\mu\nu}}{k_bT_a}\overline{M_{\mu\nu}}\,N_\nu^s\,\delta(t) = \overline{\phi_{\mu\nu}}^s\cdot\frac{1}{2}\zeta\frac{\gamma_{\mu\nu}}{k_bT_a}\,\delta(t) \qquad (C.3.14a)$$

being proportional to the stationary flux $\overline{\phi_{\mu\nu}}^s$. Furthermore this flux pulse generates a disturbance

$$-\frac{1}{2}\zeta\frac{\gamma_{\mu\nu}}{k_bT_a}\overline{\phi_{\mu\nu}}^s$$

of N_ν at t=0 and a disturbance of opposite sign of N_μ. In total the disturbance of N_μ at t=0 by the voltage pulse (C.3.10) is given by summation over all flux pulses including the state μ. With $\gamma_{\mu\nu} = -\gamma_{\nu\mu}$ we get

$$(N_\mu(o) - N_\mu^S) = \frac{1}{2}\zeta\sum_{\nu=1}^{m}\gamma_{\mu\nu}\;(\overline{\phi_{\mu\nu}}^s + \overline{\phi_{\nu\mu}}^s) \qquad (C.3.14b)$$

This disturbance of N_μ at t=0 represents the initial condition from which the linear response j(t) for t>0 may easily be derived with the use of (C.2.7), (C.2.9), (C.2.18)-(C.2.21) and with the definition of the fundamental solutions

$$j(t>0) = \frac{1}{2}\zeta\frac{1}{k_bT_a}\sum_{K,\nu=1}^{m}\;\sum_{\mu,\rho=1}^{n}\gamma_{\mu\nu}\,\gamma_{\kappa\rho}\,M_{\kappa\rho}\,(\overline{\phi_{\mu\nu}}^s + \overline{\phi_{\nu\mu}}^s)\;\Omega_{\rho\mu}(t) \qquad (C.3.15)$$

The delta-shaped response at t=0 is determined by (C.3.14a). This yields the total response j(t):

$$j(t) = 0 \quad \text{for} \quad t < 0$$

$$j(t \geq 0) = \zeta\frac{1}{k_bT_a}\,\delta(t)\cdot\frac{1}{2}\sum_{\mu,\nu}\gamma_{\mu\nu}^{2}\,\overline{\phi_{\mu\nu}}^s + \sum_{\kappa,\mu,\nu,\rho}\gamma_{\mu\nu}\,\gamma_{\kappa\rho}\,\overline{\phi_{\mu\nu}}^s\,M_{\kappa\rho}\,\Omega_{\rho\mu}(t) \qquad (C.3.16)$$

The complex admittance function $Y(\omega)$ is according to (C.3.12) given by Fourier transformation of $j(t)$. For calculation of the spectrum $G_{\Delta J}(\omega)$ of current fluctuations at equilibrium we are interested in the real part of $Y(\omega)$:

$$4\, k_b T_a \operatorname{Re} Y(\omega) = 2 \sum_{\mu,\nu} \gamma_{\mu\nu}^{2}\, \overline{\phi_{\mu\nu}}^{s} + \qquad (C.3.17)$$

$$4 \sum_{\kappa,\mu,\nu,\rho} \gamma_{\mu\nu}\, \gamma_{\kappa\rho}\, \frac{1}{2} (\overline{\phi_{\mu\nu}}^{s} + \overline{\phi_{\nu\mu}}^{s})\, M_{\kappa\rho} \int_0^\infty \Omega_{\rho\mu}(t) \cos \omega t \, dt$$

According to the Nyquist theorem (C.3.1) by (C.3.17) the equilibrium current fluctuations in discrete systems can generally be calculated. (C.3.17) is valid if one starts from the master equation as well as in the case of negligible interactions, where the linear phenomenological equations are used as basic equations. Analogous expressions for the admittance can be derived for other transport observables T than electric current and other forces than voltage acting on the system. We emphasize that the derived expression for the admittance is not restricted to equilibrium steady states. Below we shall use the result for a comparison with our result for nonequilibrium transport fluctuations. At equilibrium according to detailed balance (C.2.8) the symmetric part $\frac{1}{2} (\overline{\phi_{\mu\nu}}^{s} + \overline{\phi_{\nu\mu}}^{s})$ of the stationary flux matrix equals $\overline{\phi_{\mu\nu}}^{s}$. Naturally the admittance (C.3.17) can alternatively be derived by calculation (in the frequency domain) of the response to a periodic force (Frehland, 1981).

4. Theory of Transport Fluctuations Around Equilibrium and Non-equilibrium Steady States

In this section we derive a general approach to steady state transport fluctuations in discrete systems. The presentation will be organized as follows: In section 4.1 we study the time correlations between individual fluxes in the system building up a fourth order correlation matrix. Then the derivation of the autocorrelation function and spectral density of the current fluctuations is done in 4.2 by simple summation over the components of this correlation matrix.
For those readers who are interested in a more rigorous mathematical treatment of the time correlations, in 4.3 a derivation is given by taking the limit of rectangular events to δ-pulse fluxes similarly as in section 2.2 of part B for the Poisson process.

4.1 The Flux-Correlation Matrix

We introduce a correlation matrix $\underset{\sim}{C}(t)$ of fourth order with the elements $C_{\mu\nu,\kappa\rho}(t)$ as the time correlations between the individual fluxes $\phi_{\mu\nu}(o)$ and $\phi_{\kappa\rho}(t)$.

$$C_{\mu\nu,\kappa\rho}(t) := \langle\phi_{\mu\nu}(o)\phi_{\kappa\rho}(t)\rangle - \langle\phi_{\mu\nu}^{s}\rangle\langle\phi_{\kappa\rho}^{s}\rangle \qquad \text{(C.4.1)}$$

The second term on the right-hand side is included in the definition of $\underset{\sim}{C}$ because we are interested in only the fluctuating part of the transport properties. The correlations can be referred to time zero without loss of generality, as steady states are considered.

$\underset{\sim}{C}(t)$ is essentially determined by the fundamental solution matrix $\underset{\sim}{\Omega}$:

$$C_{\mu\nu,\kappa\rho}(t) = \phi_{\mu\nu}{}^{s} \left[\delta_{\mu\nu,\kappa\rho} \cdot \delta(t) + M_{\kappa\rho}\, \Omega_{\rho\mu}(t) \right] \qquad (C.4.2)$$

with

$$\delta_{\mu\nu,\kappa\rho} = \begin{cases} 1 & \text{for } \mu,\nu = \kappa,\rho \\ 0 & \text{else} \end{cases}$$

The validity of (C.4.2), which is the basic fundamental relation for the treatment of transport fluctuations in discrete transport systems, may be seen by the following arguments: Taking the ensemble average at time zero nonvanishing contributions to $C_{\mu\nu,\kappa\rho}(t)$ are possible only from those realizations, where at t=0 a transition $\nu\to\mu$ occurs, if we are starting from the master equation (C.2.1), or a particle jumps from place ν to place μ, if we use the linear phenomenological equations (C.2.16). Thus,the ensemble average in (C.4.1) is a conditional average, which has to be taken only over the subensemble given by the 'initial condition', that the system (particle) is in state μ (place μ) at t=0. The mean rate of transitions (jumps) $\nu\to\mu$ is equal to the steady state flux $\phi_{\mu\nu}{}^{s}$. The δ-like first term in (C.4.2) is the correlation of transitions (jumps) $\nu\to\mu$ with themselves at t=0. By the 'initial condition', that the system at t=0 is in state μ, the fundamental solution $\Omega_{\rho\mu}(t)$ appears in the second time dependent term in (C.4.2).

The relation (C.4.2) is valid for $t\geq 0$. For derivation of the autocorrelation function $C_{\Delta J}(t)$ e.g. of current fluctuations this is sufficient to derive $C_{\Delta J}(t\geq 0)$ and because of the time symmetry (A.4.2b) of C in stationary processes $C_{\Delta J}(t\leq 0)$ is then determined as well. Nevertheless the correlation matrix may easily be derived from (C.4.2). From the definition (C.4.1) follows immediately

$$C_{\mu\nu,\kappa\rho}(t) = C_{\kappa\rho,\mu\nu}(-t) \tag{C.4.3}$$

because a shift in time for stationary processes is irrelevant:

$$<\phi_{\mu\nu}(o)\phi_{\kappa\rho}(t)> = <\phi_{\mu\nu}(-t)\ \phi_{\kappa\rho}(o)>$$

4.2 Autocorrelation Function and Spectral Density of Current Fluctuations

The autocorrelation function of current J (or some other transport observable T) is with (C.2.9) or (C.2.21)

$$C_{\Delta J}(t) = <\Delta J(o)\,\Delta J(t)> = \sum_{\mu,\nu,\kappa,\rho} \gamma_{\mu\nu}\ \gamma_{\kappa\rho}\ C_{\mu\nu,\kappa\rho}(t)$$

$$= \sum_{\mu,\nu} \delta(t)\,\gamma_{\mu\nu}{}^2\ \phi_{\mu\nu}{}^S + \sum_{\mu,\nu,\kappa,\rho} \gamma_{\mu\nu}\ \gamma_{\kappa\rho}\ \phi_{\mu\nu}{}^S\ M_{\kappa\rho}\ \Omega_{\rho\mu}(t) \tag{C.4.4}$$

and the spectral density is given by the Wiener-Khintchine relations (B.1.7)

$$G_{\Delta J}(\omega) = 2\sum_{\mu,\nu} \gamma_{\mu\nu}{}^2\ \phi_{\mu\nu}{}^S \tag{C.4.5}$$
$$+ 4\sum_{\mu,\nu,\kappa,\rho} \gamma_{\mu\nu}\ \gamma_{\kappa\rho}\ \phi_{\mu\nu}{}^S\ M_{\kappa\rho} \int_0^\infty \Omega_{\rho\mu}(t)\ \cos\omega t\ dt$$

Because we will discuss properties of the steady state fluctuations especially with the use of (C.4.5) under general aspects and for a number of special transport models, in this section we want to make only few short remarks. The spectral density consists of two terms:

The first one is white noise and is generated because of the discrete structure of the transport process similarly as shot noise (c.f. sections B.2.3 and C.1.2). The second term comes from the time dependent behaviour of the system via the fundamental solutions of the master equation, the linear phenomenological equation respectively. A simple first comparison with the structure of the real part of the admittance (C.3.17) shows immediately equality in the equilibrium situation, because in this case the stationary flux matrix is symmetric according to the detailed balance (C.2.8). At nonequilibrium the asymmetric parts of $\underset{\sim}{\phi}^{s}$ yield an additional contribution to the fluctuations . The following decomposition of spectral density is derived from the decomposition of $\underset{\sim}{\phi}^{s}$ into its symmetric and antisymmetric parts and using (C.3.17).

$$G_{\Delta J}(\omega) = 4\, k_b T_a \operatorname{Re} Y(\omega) + 4 \sum_{\mu,\nu,\kappa,\rho} \gamma_{\mu\nu}\, \gamma_{\kappa\rho}\, \frac{1}{2}\, (\phi_{\mu\nu}{}^{s} - \phi_{\nu\mu}{}^{s})\, M_{\kappa\rho} \int_{0}^{\infty} \Omega_{\rho\mu}(t) \cos \omega t \, dt \tag{C.4.6}$$

Thus at nonequilibrium a second term arises which expresses a characteristic difference between equilibrium and nonequilibrium fluctuations. Temperature T_a comes in through the voltage-dependence (C.3.5) of the rate constants (i.e. through the external force!) but not in the fluctuations (see (C.4.2)-(C.4.5)). The transport system can be assumed to be in a bath of defined temperature.The problem of temperature at nonequilibrium will not further be discussed.

4.3 More Rigorous Treatment of the Time Correlation Matrix

We now give a derivation of the central relation(C.4.2) for the time correlation matrix, which is more satisfying from the mathematical point of view. As in the treatment of Poisson pulse sequences in section B.2.2 we start with rectangular pulses and take the limit to δ-pulses. We take the master equation as basic equation describing the probabilistic behaviour of the fluxes in the limit to δ-pulses. In case the linear phenomenological equations are used, the derivation can be done analogously. We make the assumptions that 1. pulse-height h and -duration τ are chosen such that

$$h \ \tau = 1 \tag{C.4.7}$$

remains constant also in the limit $h \to \infty$, $\tau \to 0$, and 2. for sufficiently small τ overlapping of pulses can be neglected. Then for finite h and τ the flux $\phi_{\mu\nu}$ is a sequence of rectangular pulses and $\phi_{\mu\nu}$ can take the values h or 0.

Introducing the (time dependent) mean rates $\lambda_{\mu\nu}(t)$ of pulses, the probabilities of $\phi_{\mu\nu}$ to take the values h or 0 are (overlapping neglected!)

$$P(\phi_{\mu\nu} = h, \ t) = \lambda_{\mu\nu}(t) \cdot \tau$$

$$P(\phi_{\mu\nu} = 0, \ t) = 1 - \lambda_{\mu\nu}(t) \cdot \tau \tag{C.4.8}$$

In the limit $h \to \infty$, $\tau \to 0$ the mean rates $\lambda_{\mu\nu}$ are assumed to be helt constant; then they are simply given by the solutions P_ν of the master equation (C.2.1) according to (C.2.6)

$$\lambda_{\mu\nu}(t) = M_{\mu\nu} \ P_\nu(t) = \langle\phi_{\mu\nu}(t)\rangle \tag{C.4.9}$$

The expectation value $<\phi_{\mu\nu}(t)>$ does not change during the limiting process, because $<\phi_{\mu\nu}>$ is according to the definition (A.2.7) of expectation values with (C.4.8)

$$<\phi_{\mu\nu}(t)> = h \cdot P(\phi_{\mu\nu}=h,t) = h\, \lambda_{\mu\nu}(t) \cdot \tau = \lambda_{\mu\nu}(t) \qquad (C.4.10)$$

With (C.4.8) and because the fluxes can take only the values h and 0 the time correlations (t>o) between different fluxes are

$$<\phi_{\mu\nu}(o)\ \phi_{\kappa\rho}(t)> = h^2\ P(\phi_{\mu\nu}=h,o;\ \phi_{\kappa\rho}=h,t) \quad , \quad \mu,\nu \neq \kappa,\rho \qquad (C.4.11)$$

The joint probability in (C.4.10) can be expressed by the conditional probability according to (A.2.8)

$$P(0,\phi_{\mu\nu}=h;t,\phi_{\kappa\rho}= h) = P(\phi_{\kappa\rho}=h,t/ \ \phi_{\mu\nu}=h,0)\cdot P(\phi_{\mu\nu}=h,0) \qquad (C.4.12)$$

With (C.4.8) the probabilities can be replaced by the corresponding mean rates and (C.4.11) yields

$$<\phi_{\mu\nu}(o)\ \phi_{\kappa\rho}(t)> = h^2 \cdot \lambda_{\kappa\rho}(t)_{\,(\phi_{\mu\nu}(o)=h)} \cdot \tau\cdot\lambda_{\mu\nu}(o)\cdot\tau$$

$$= \lambda_{\mu\nu}(o)\ \lambda_{\kappa\rho}(t)_{\,(\phi_{\mu\nu}(o)=h)} \qquad (C.4.13)$$

Now we use (C.4.9). Then $\lambda_{\mu\nu}(o)$ is the stationary flux $\phi_{\mu\nu}^{\ s}$. In the limit $h\to\infty$, $\tau\to 0$ the conditional mean rate $\lambda_{\kappa\rho}$ in (C.4.13) is the expected flux under the condition that at t=0 a transition $\nu\to\mu$ occurs, i.e. the system at t=0 is in state μ. With the definition of the fundamental solution $\Omega_{\mu\nu}(t)$ to be the deviation from steady state $P_\nu^{\ s}$ under just this initial condition, we get in the limit $h\to\infty$, $\tau\to 0$ for $t\geq 0$ and $\mu,\nu\neq\kappa,\rho$

$$<\phi_{\mu\nu}(0)\ \phi_{\kappa\rho}(t)> = \phi_{\mu\nu}{}^{s}(M_{\kappa\rho}\ \Omega_{\rho\mu}(t) + \phi_{\kappa\rho}{}^{s})$$

$$= \phi_{\mu\nu}{}^{s}\ M_{\kappa\rho}\ \Omega_{\rho\mu}(t) + \phi_{\mu\nu}{}^{s}\ \phi_{\kappa\rho}{}^{s} \qquad (C.4.14)$$

For $\mu,\nu=\kappa,\rho$ we still have to take into account the correlation of pulses with themselves, which can be treated as done in B.2.2 for Poisson sequences and yields a δ-like contribution (times the mean rate $\phi_{\mu\nu}{}^{s}$). Hence we generally get from (C.4.13) and (B.2.13b)

$$<\phi_{\mu\nu}(o)\ \phi_{\kappa\rho}(t)> - \phi_{\mu\nu}{}^{s}\ \phi_{\kappa\rho}{}^{s} = \phi_{\mu\nu}{}^{s}\ \delta_{\mu\nu,\kappa\rho} \cdot \delta(t)$$

$$+ \phi_{\mu\nu}{}^{s}\ M_{\kappa\rho}\ \Omega_{\rho\mu}(t), \qquad (C.4.15)$$

in accordance with (C.4.2).

5. Transport Fluctuations in Basic Membrane Transport Systems

Two basic concepts of ion transport through membranes are the concepts of facilitated ion transport through hydrophilic pathways (pores or channels) and of transport mediated by carrier molecules. As first applications we now discuss both transport models. Recently the presented theoretical approach has been applied to a further simple mechanism, the transnort of hydrophobic ions across lipid membranes (Junges, Kolb, 1982 , Jordan, 1980).

5.1 Current Noise in Open Channels

We apply the theory of the preceding section to the one-dimensional model of ion transport through pores, starting from the linear phenomenological equation (C.2.16). The simple example of pores with one binding site has been discussed in section 1. The transport system (Läuger (1973), Frehland, Läuger (1974), Frehland (1978)) consists of a membrane separating two ionic solutions and containing a great number of identical pores (channels). It is assumed that only one ion species may penetrate the channels. The concentrations of the ions at the pore mouths in the solutions are assumed to be helt constant and act as reservoirs. The pores are considered to be a sequence of (n+1) activation barriers separated by n energy minima (binding sites).

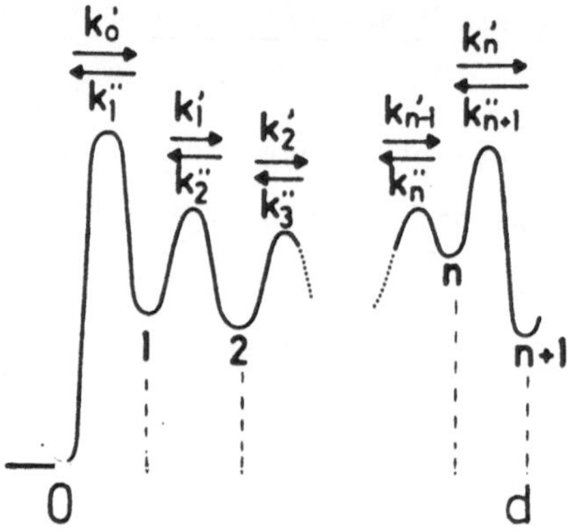

Fig. C1:

Potential profile in ionic channels with n-binding sites within the channel, 0 and (n+1) denoting the left-, right-hand reservoirs respectively.

The rate constants for jumps from the i-th binding site to the right or to the left are denoted by k_i', k_i'' respectively (see Fig. C1). Jumps are possible only between neighbouring sites. As consequence the matrix $\underset{\sim}{M}$ of coefficients in the phenomenological equations (C.2.16) has a simple tridiagonal structure

$$M_{ii} = -(k_i' + k_i''),$$
$$M_{i,i+1} = k_{i+1}'' \quad , \quad M_{i+1,i} = k_i' \tag{C.5.1}$$
$$M_{ik} = 0 \text{ otherwise}$$

and the inhomogeneity $\underset{\sim}{Y}$ is

$$Y_1 = k_o' \, N_o \quad , \quad Y_n = k_{n+1}'' \, N_{n+1}$$
$$Y_i = 0 \text{ otherwise.} \tag{C.5.2}$$

N_o, N_{n+1} are the constant concentrations at the pore mouths and k'_o, k''_{n+1} the rate constants for jumps into the pores from the left, right solutions respectively. It follows immediately from the special tridiagonal structure of M that $\underset{\sim}{M}$ may be symmetrized and the eigenvalues λ_i are real and negative. The symmetrization is done by a diagonal matrix $\underset{\sim}{T}$ with the non-vanishing components

$$T_{ii} = t_1 \cdot t_2 \cdot \ldots \cdot t_{i-1} \cdot t_i$$

$$t_1 = 1, \; t_i = \sqrt{k''_i / k'_{i-1}} \qquad \text{for } i = 2,3,\ldots,n \qquad (C.5.3)$$

Because $\underset{\sim}{M}$ may be symmetrized, it can be diagonalized to $\underset{\sim}{\hat{M}}$ by an appropriate matrix $\underset{\sim}{V}$, which is constructed by standard procedures. The fundamental solutions are then given by

$$\Omega_{ik}(t) = \sum_{j=1}^{n} V^{-1}_{ij} \, e^{-\lambda_j t} \, V_{jk} \qquad (C.5.4)$$

Because the transport model is one-dimensional and jumps are allowed to adjacent sites only, the non-vanishing components of the flux matrix are $\phi_{i+1,i}$ and $\phi_{i-1,i}$. Therefore it is useful to introduce the unidirectional fluxes over the i-th barrier (c.f. section 1.2)

$$\phi'_i = \phi_{i,i-1} \quad , \quad \phi''_i = \phi_{i-1,i} \qquad (C.5.5$$

and the total flux ϕ_i over the i-th barrier

$$\phi_i = \phi'_i - \phi''_i \quad , \quad i = 1,2,\ldots,n+1 \qquad (C.5.6)$$

The electric current in the simplified notation is a linear combination of the ϕ_i

$$J = \sum_{i=1}^{n+1} \gamma_i \, \phi_i \qquad (C.5.6)$$

where the γ_μ take into account the contribution to the electric current from the jumps over the i-th barrier. They may depend on geometric properties (e.g. distances between the binding sites) and dielectric properties of the membrane transport system. With the notations (C.5.5) and (C.5.6) the fourth-order correlation matrix $\underset{\sim}{C}$ may be replaced by a second-order matrix with the elements $<\phi_i(0)\ \phi_j(t)>$, which are with the use of (C.4.2)

$$<\phi_i(0)\ \phi_j(t)> = \phi_i^s\ \phi_j^s + \delta_{ik}(k'_{i-1}\ N^s_{i-1} + k''_i\ N^s_i)\ \delta(t)$$

$$+ \Big[k'_{i-1}\ N^s_{i-1}(k'_{k-1}\ \Omega_{k-1,i} - k''_k \Omega_{ki}) \qquad (C.5.7)$$

$$+ k''_i\ N^s_i(k''_k \Omega_{k,i-1} - k'_{k-1}\Omega_{k-1,i-1})\Big]$$

In (C.5.7) Ω_{ok}, Ω_{ko}, $\Omega_{n+1,k}$, $\Omega_{k,n+1}$ are to be set zero. Summation over $<\phi_i(0)\phi_k(t)>$ yields the autocorrelation function of current fluctuations, which reads after some manipulations by changing the summation indices:

$$C_{\Delta J}(t) = \sum_{i=1}^{n+1} \gamma_i^2(k'_{i-1}\ N^s_{i-1} + k''_i\ N^s_i)\ \delta(t)$$

$$+ \sum_{i,k=1}^{n} \Omega_{ki}(t)\ \Big[k'_{i-1}\ N^s_{i-1}\ (\gamma_i\gamma_{k+1}\cdot k' - \gamma_i\gamma_k \cdot k'')$$

$$+ k''_{i+1}\ N^s_{i+1}\ (\gamma_{i+1}\ \gamma_k \cdot k'' - \gamma_{i+1}\ \gamma_{k+1}\cdot k')\Big] \qquad (C.5.8)$$

The spectral density $G_{\Delta J}(\omega)$ is given by the Wiener-Khintchine relations (B.1.7)

$$G_{\Delta J}(\omega) = 2 \sum_{i=1}^{n+1} \gamma_i^2 \, (k'_{i-1} \, N^s_{i-1} + k''_i \, N^s_i)$$

$$+ 4 \sum_{l=1}^{n} \tau_l \, (1+\omega^2 \tau_l^2)^{-1} \sum_{i,k=1}^{n} V^{-1}_{kl} \, V_{li}$$

$$\times \Big[k'_{i-1} \, N^s_{i-1} \, (\gamma_i \gamma_{k+1} \cdot k'_k - \gamma_i \gamma_i \cdot k''_k)$$

$$+ \; k''_{i+1} \, N^s_{i+1} \, (\gamma_{i+1} \gamma_k \cdot k''_k - \gamma_{i+1} \gamma_{k+1} \cdot k'_k) \Big] \qquad \text{(C.5.9)}$$

where we have introduced the relaxation times

$$\frac{1}{\tau_l} = \lambda_l \qquad \text{(C.5.9a)}$$

Because all eigenvalues are real, the spectrum consists of a sum of Lorentzian terms and a white noise term dominating for high frequencies.

For comparison the real part of the admittance follows from (C.3.17) with (C.5.1), (C.5.2) to be

$$4 \, k_b T_a \, \mathrm{Re} \, Y(\omega) = 2 \sum_{i=1}^{n+1} \gamma_i^2 \, (k'_{i-1} \, N^s_{i-1} + k''_i \, N^s_i)$$

$$+ 2 \sum_{l=1}^{n} \tau_l \, (1+\omega^2 \tau_l^2)^{-1} \sum_{i,k=1}^{n} V^{-1}_{kl} \, V_{li}$$

$$\times \Big[(k'_{i-1} \, N^s_{i-1} + k''_i \, N^s_i) \, (\gamma_i \gamma_{k+1} \cdot k'_k - \gamma_i \gamma_k \cdot k''_k)$$

$$+ \; (k'_i \, N^s_i + k''_{i+1} \, N^s_{i+1}) \, (\gamma_{i+1} \gamma_k \cdot k''_k - \gamma_{i+1} \gamma_{k+1} \cdot k'_k) \Big]$$

$$\text{(C.5.10)}$$

5.2 The Influence of Voltage and Barrier Structure on the Fluctuations

In a recent paper (Frehland, Faulhaber, 1980) we have discussed by numerical calculations the influence of voltage and barrier structure of the channels on the transport properties. It turned out that the influence of the internal barrier structure especially on the current fluctuation properties can be of great complexity. Nevertheless we now want to present some of these calculations which we hope to lead to a better understanding especially of nonequilibrium transport fluctuations in systems with a complex internal structure.

The calculations have been done by numerical evaluation of the eigenvalues and eigenvectors of the matrix of coefficients (C.5.1) in the transport equations. We have chosen model pores with six internal binding sites (n=6). In the figures C2-C6 the chosen values of the rate constants are given in the legends. The constant concentration at the pore mouths are set equal to one. For simplicity all γ_i, i.e. the contributions of individual jumps to the measured current, are set equal.

The potential profile (barrier structure) is indicated in all figures.

In Fig. C2 the influence of the height of an internal barrier on the equilibrium fluctuations is demonstrated. At equilibrium according to the fluctuation dissipation theorem (C.3.1) or (C.5.9) and (C.5.10) the spectral density $G_{\Delta J}(\omega)$ of current fluctuations equals 4 kT times the real part of the admittance. The results confirm general properties of transport noise at equilibrium: in general the low frequency (white noise) limit of spectral density is lower than the high frequency limit (inverse Lorentzian behaviour, Kolb and Läuger (1977), Läuger (1978), Frehland (1978), Frehland and Stephan

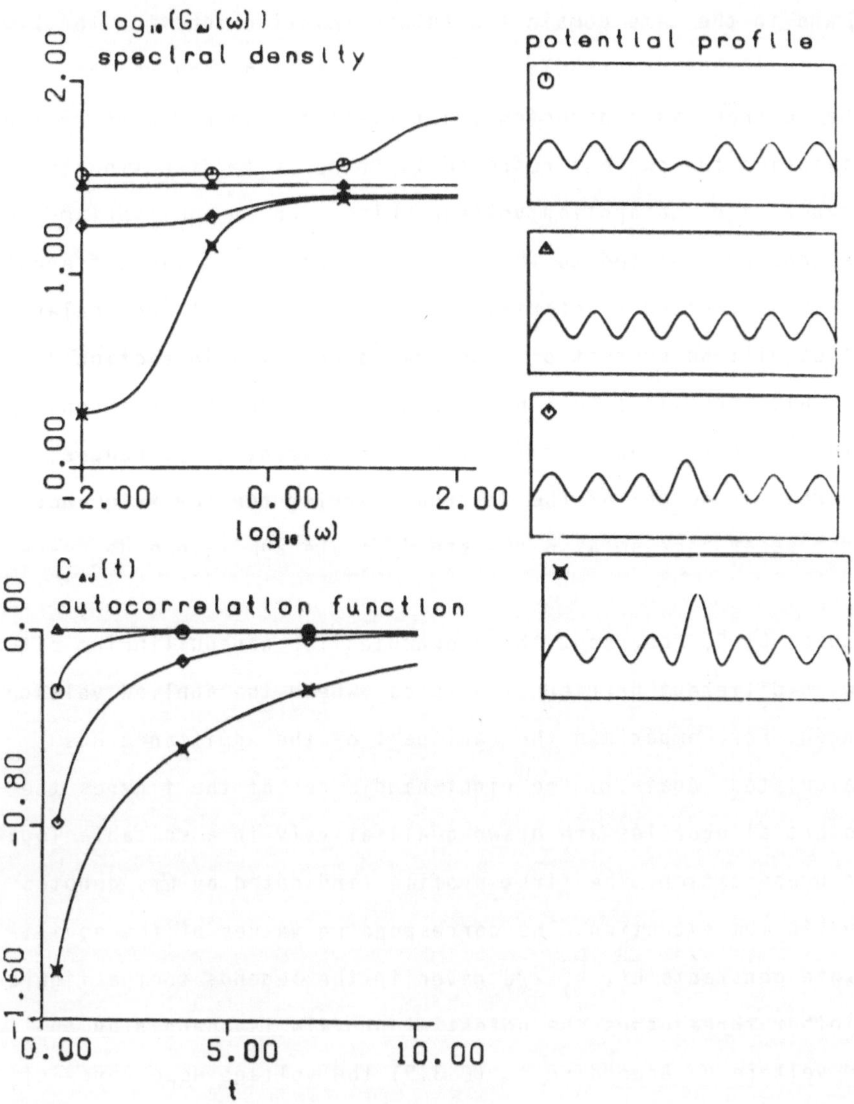

Fig. C2: Equilibrium current fluctuations in pores: variation of the height of the central barrier. $n=6$, $N_0=N_{n+1}=1$, $k_i'=k_i''=1$ for $i=0,1,\ldots,n+1$ with the exception $k_3'=k_4''=$ { ○ 10; △ 1; ◇ 0.2; ✱ 0.01 }

All γ_i are chosen to be equal.

(1979)) and in the time domain the autocorrelation function for $t>0$ is negative (Frehland, 1980). This is a direct consequence of the fact that current is a directed (vectorial) transport quantity and at equilibrium the current response is in opposite direction to a disturbance, e.g. an applied voltage pulse. And at equilibrium the fluctuations are related to the response properties. The different behaviour of directed (vectorial) quantities (fluxes) and scalar quantities will be subject of a general discussion in section 7.
For a regular profile (equal barrier heights) the spectral density and autocorrelation function for $t>0$ are constant (c.f. Läuger, 1978). With increasing height of the internal barrier the low frequency limit decreases, because also the steady state admittance decreases.

The figures C3-C6 represent three examples for nonequilibrium situations with different barrier structures, where the applied voltage is changed. For comparison the real part of the admittance has been calculated. Again, on the right-hand sides of the figures the four potential profiles are drawn qualitatively in a suitable logarithmic presentation. The first profile (indicated by ⓪) denotes the equilibrium situation. The corresponding values of the equilibrium rate constants $\overline{k_i'}$, $\overline{k_i''}$ are given in the legends to the figures. In the other three cases the potential profile is changed by an applied voltage u. According to (C.3.5) the voltage dependence of the rate constants is given by

$$k_i' = \overline{k_i'}\, e^{-\alpha_{i+1}\frac{u}{2}} \qquad , \quad k_i'' = \overline{k_i''}\, e^{-\alpha_i \frac{u}{2}}$$

$$u = \frac{ze_oV}{kT}, \quad \alpha_i\, ze_o = \gamma_i \qquad \text{(C.5.11)}$$

In Fig. C3 the low frequency difference (the second term in (C.4.6)) between $G_{\Delta J}(\omega)$ and 4 kT Re $Y(\omega)$ increases with increasing voltage. This difference ('excess noise') is usual Lorentzian noise. It has an intensity, which is approximately proportional to the square of the applied voltage.

In Fig. C4 the influence of a low barrier within the pore is investigated, favouring the stay of ions in the central part of the pores. For low frequencies a comparison of Re $Y(\omega)$ with $G_{\Delta J}(\omega)$ again demonstrates the occurrence of typical Lorentzian excess noise. But on the contrary to the results for a regular barrier structure, the spectral density at nonequilibrium assumes a strong minimum. The interpretation is as follows: The fluctuations consist of a sort of superposition of the global transport through the pore, being responsible for the 'excess' fluctuations, and the transport fluctuations generated by the ionic jumps (in both directions) over the low barrier in the centre of the pores. The latter contribution to noise is similar to the transport noise generated by the movement of hydrophobic ions in a similar potential profile (Kolb, Läuger, 1977). Again as in the equilibrium examples this is a direct consequence of negative correlation of vectorial fluxes. Furthermore, the increase of the frequency of ionic jumps by the lowered central barrier increases the white noise high frequency limit compared with the case of a regular barrier structure.

In Fig. C5 the influence of high barriers regulating the influx into and efflux out of the pores is investigated. In this case the excess noise, i.e. the difference between 4 kT Re $Y(\omega)$ and $G_{\Delta J}(\omega)$ for low frequer

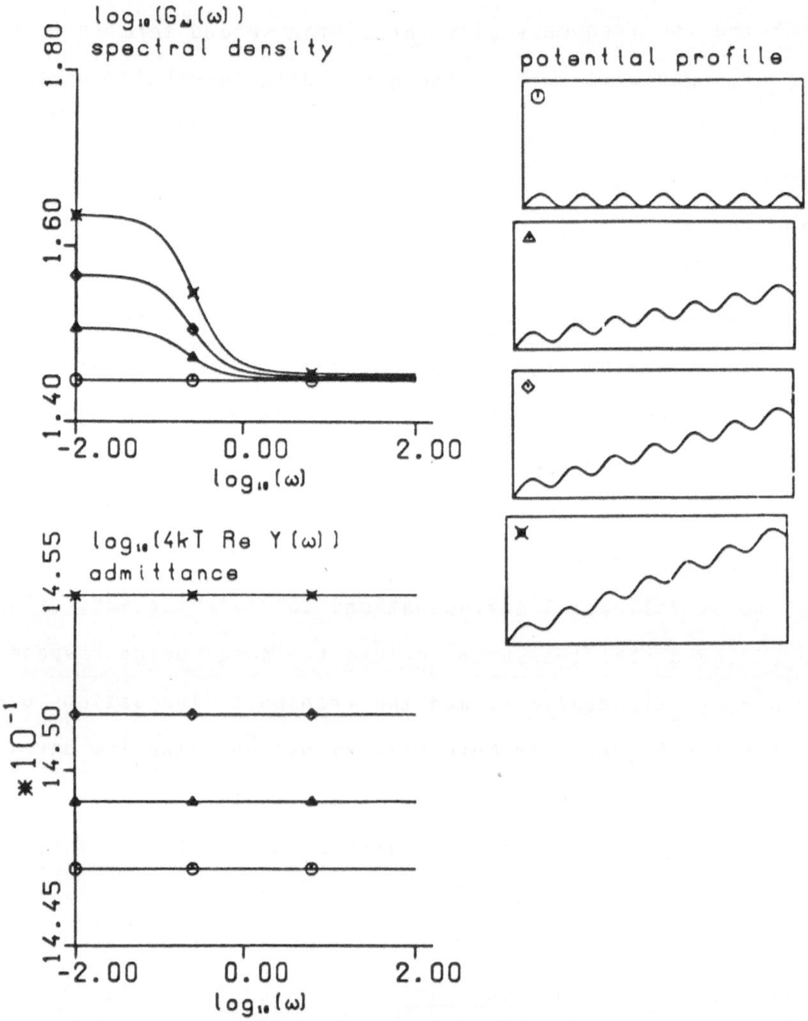

Fig. C3: Admittance and spectral density of current fluctuations. Regular barrier structure. $n=6$, $N_0=N_{n+1}=1$; all γ_i equal; $k_i'=k_i''=1$ for all $i=0,1,2,...,n+1$. Voltage dependence of the rate constants: $k_i'=k_i'e^{-\frac{u/2}{n+1}}$ $k_i''=k_i''e^{\frac{u/2}{n+1}}$. Applied (reduced) voltages u: $u=0$, $u=4/3$, $u=2$, $u=8/3$.

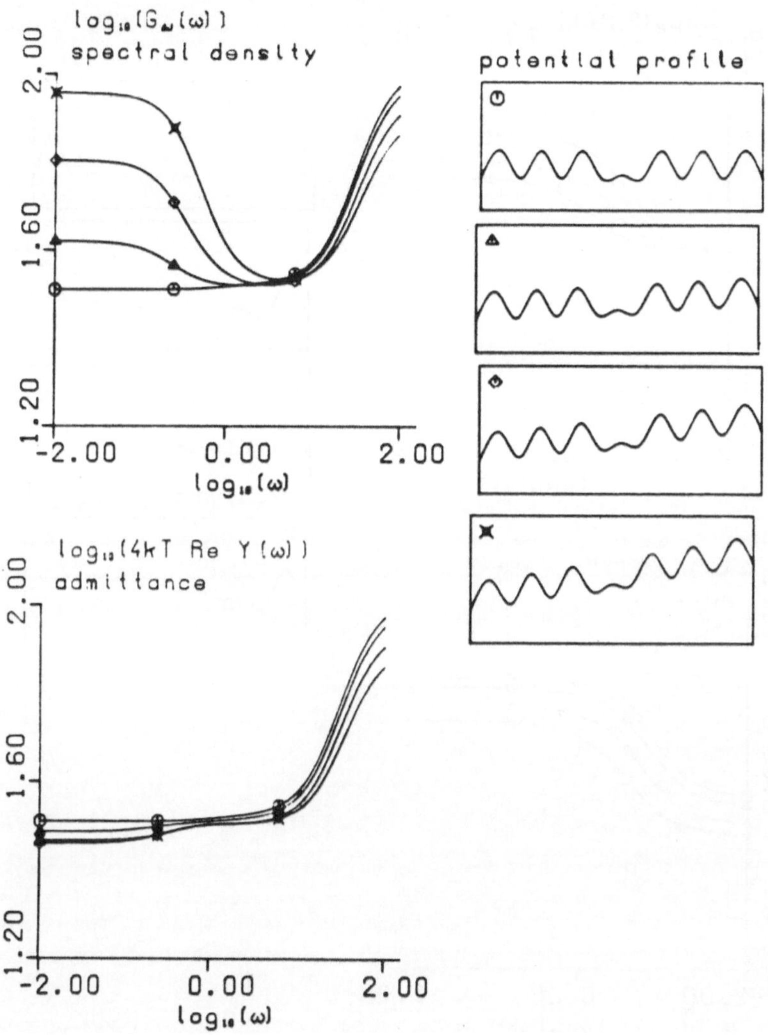

<u>Fig. C4:</u> Non-equilibrium: the effect of a low central barrier.

$n=6$, $N_0=N_{n+1}=1$; all γ_i equal; $k_i'=k_i''=1$ with the exceptions $k_3'=k_4''=20$.

Applied (reduced) voltages u: u=0, u=2, u=4, u=6.

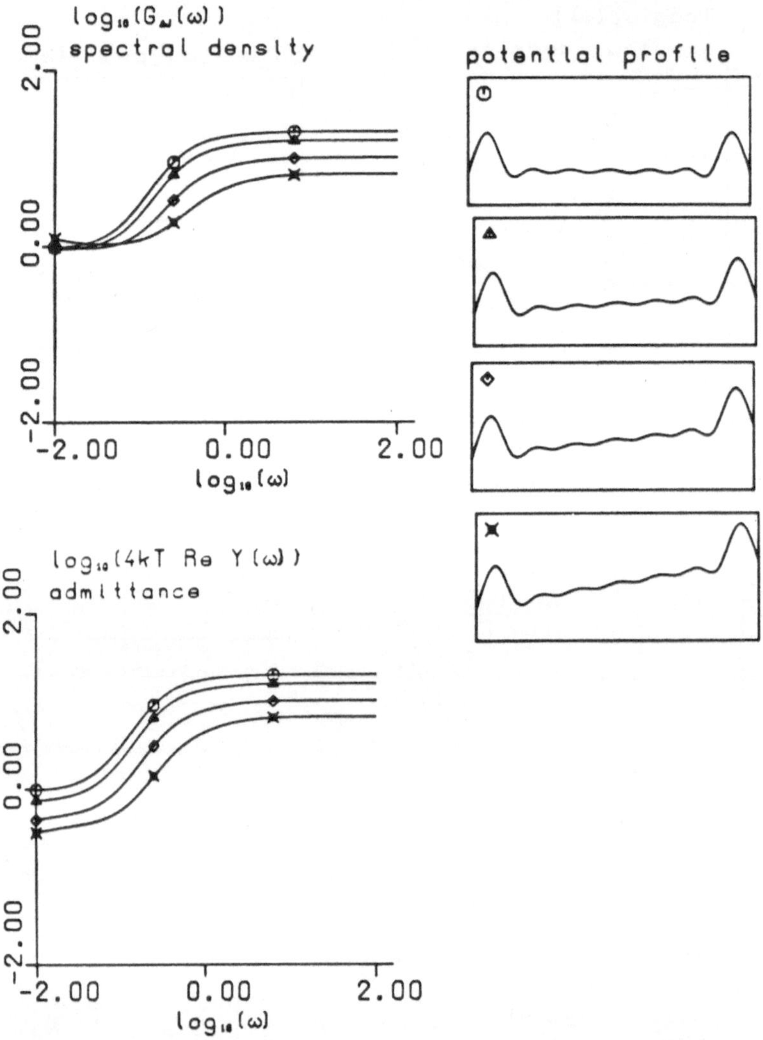

Fig. C5: Non-equilibrium: the effect of high decentral barriers, n=6, $N_0=N_{n+1}=1$; all γ_i equal; $k_i'=k_i''=1$ with the exceptions $k_0'=k_1''=k_6'=k_7''=0.02$. Applied voltages u: u=0, u=2, u=4, u=6.

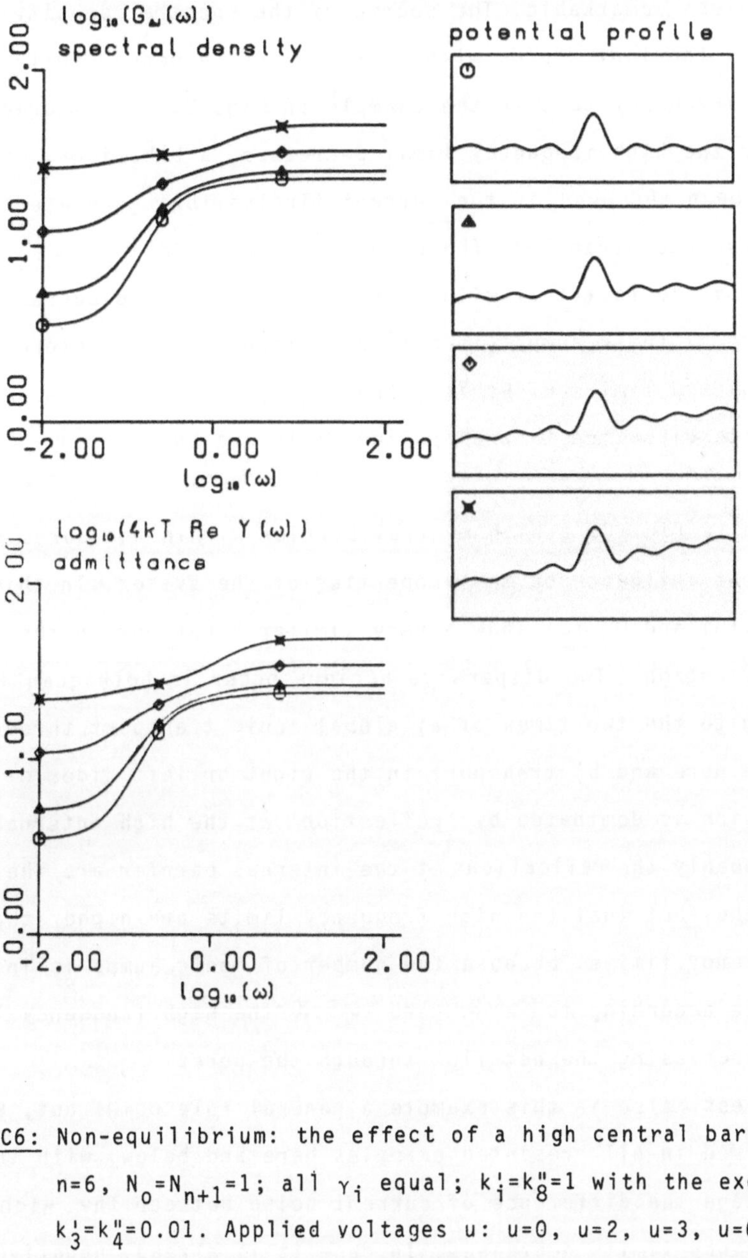

Fig. C6: Non-equilibrium: the effect of a high central barrier. $n=6$, $N_0=N_{n+1}=1$; all γ_i equal; $k_i'=k_8''=1$ with the exceptions $k_3'=k_4''=0.01$. Applied voltages u: u=0, u=2, u=3, u=6.

cies, is less remarkable. The course of the spectral density is dominated by the ionic jumps within the pores and hence similar to the high frequency part in the example in Fig. C4. For increasing voltage u the high frequency limit decreases, a behaviour, which has also been observed in the current fluctuations generated by carrier mediated transport (Kolb and Frehland, 1980). It is a consequence of the fact that with increasing voltage and current the total rate of ionic jumps (in both directions), which determine the high frequency limits of Re $Y(\omega)$ and $G_{\Delta J}(\omega)$, decreases because jumps in opposite direction to the applied field are suppressed.

According to Fig. C6 a high central barrier within the pores has a different influence on the properties of the system. In this case 4 kT Re $Y(\omega)$ and $G_{\Delta J}(\omega)$ show a very similar behaviour in the whole frequency region. Two dispersive regions occur in both quantities belonging to the two times of a) global ionic transport through the whole pore and b) transport in the right or left sides of the pores, which is dominated by 'reflection' at the high internal barrier. Probably the reflections at the internal barrier are the reason for the fact that the high frequency limits are higher than the low frequency limits, because the number of ionic jumps is increased (and hence according to (C.5.9) and (C.5.10) the high frequency limits, without increasing the net flux through the pores.
Nevertheless, also in this example a general rule comes out, which is confirmed in all presented examples here and below: with increasing voltage the difference of current noise between the high and low frequency limits decreases. In case it is already negative, its absolute value increases.

The presented numerical results may be useful to demonstrate the great complexity of the behaviour of transport systems as a function of their internal (fine) structure. These results may serve as a warning against the overinterpretation of and too specific conclusions from experimental results concerning problems as the barrier structure inside the pores. We have great doubts on such speculations. On the other hand the great variety of the fluctuation properties (e.g. expressed in the result for the spectral density) demonstrates the strong influence of the internal fine structure of transport systems on the strength of the fluctuations (in different frequency regimes). During the evolution nature might have used these possibilities for an optimization of the transport system, e.g. the minimization of noise. In this section has been discussed a very simple transport model of pores without any interactions. In case interactions are included e.g. in the single-file model of narrow pores,which will be discussed in section 6, the variety and complexity of the dependences of transport properties on internal parameters increases. E.g. it seems likely that by the single-file mechanism the current fluctuations can drastically be reduced, a point which also will be discussed below. A detailed discussion and comparison with shot noise will be done for high applied voltage.

5.3 Carrier Noise

An alternative basic concept of membrane transport is the concept of carrier mediated ion transport. E.g. the valinomycin-mediated transport through lipid bilayers has been extensively investigated (Stark et al. (1971), Läuger and Stark (1971), Knoll and Stark (1975). We briefly describe the analysis of carrier current noise in nonequilibrium steady states. The experimental and theoretical analysis of carrier noise has recently been published (Kolb and Frehland, 1980) and confirmed our theory of transport fluctuations presented in section 4.

The mechanism as proposed by Läuger and Stark (1971) and Stark et al. (1971) represents an important example, which shows the applicability and usefulness of the discrete transport concept. It is assumed that the single carrier molecules act independently. As shown in Fig.C7 the transport takes place in four steps:
a) recombination of ion M^+ and neutral carrier at the left-hand interface ('), b) translocation of the complex to the right-hand interface ("), c) dissociation of the complex and release of the ion into the solution and d) back transport of the free carrier.

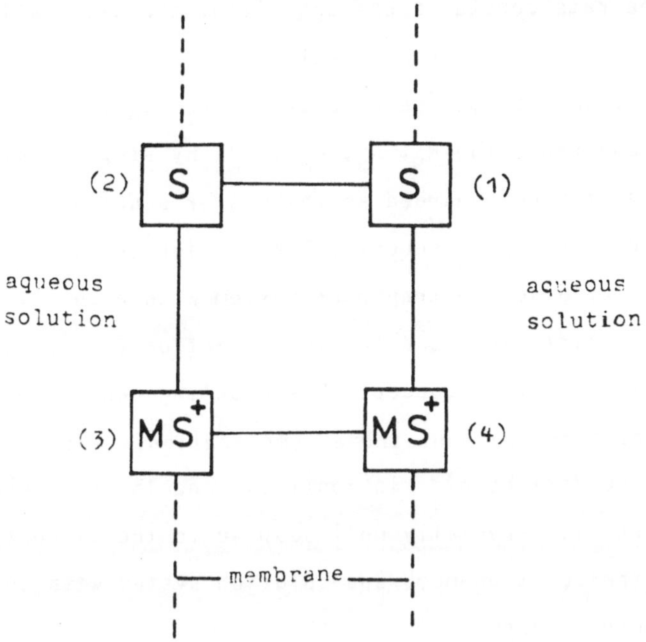

Fig. C7: State diagram for the simplest model of carrier-mediated ion transport

The total carrier concentration N_o in the membrane is constant

$$N_o = N'_S + N''_S + N'_{MS} + N''_{MS} \qquad (C.5.12)$$

where N'_S, N''_S, N'_{MS}, N''_{MS} are the interfacial concentrations at the left (') and right (") interface, respectively. The time-dependent behaviour of the system is given by the following equations:

$$dN'_S/dt = -k_R C_M N'_S + k_D N'_{MS} - k_S(N'_S - N''_S),$$

$$dN''_S/dt = -k_R C_M N''_S + k_D N''_{MS} - k_S(N''_S - N'_S),$$

$$dN'_{MS}/dt = k_R C_M N'_S - k_D N'_{MS} - k'_{MS} N'_{MS} + k''_{MS} N''_{MS},$$

$$dN'_{MS}/dt = k_R C_M N''_S - k_D N''_{MS} - k''_{MS} N''_{MS} + k'_{MS} N'_{MS}. \qquad (C.5.13)$$

k_R, k_D are the rate constants characterizing the association, dissociation respectively. The translocation of the uncharged, charged carrier is described by the rate constants k_S, k'_{MS}, k''_{MS}, respectively. Setting $N_0=1$ and replacing N'_S, N''_S, N'_{MS}, N''_{MS} by $P(N'_S)$, $P(N''_S)$, $P(N'_{MS})$, $P(N''_{MS})$ (C.5.13) may be regarded as the master equation for one carrier as a single transport subunit. The carrier system represents a simple example of discrete transport systems with a coupling between transport (jump diffusion) and chemical reaction (dissociation-association). In the discrete concept this coupling can be treated within the kinetic equations without those complications, which arise from boundary conditions etc. in continuum models. Furthermore, the discrete concept is especially well-adapted to the discontinuous structure of the solution-membrane-solution system with an extension on the microscopic range.

The application of the theory described in section 4 yields by very lengthy calculations.which have been described (Kolb and Frehland, 1980) the following result for the spectral density $G_{\Delta J}(\omega)$.

$$G_{\Delta J}(\omega) = 2z^2e_0^2 \times \left(k'_{MS}\,\overline{N'_{MS}} + k''_{MS}\,\overline{N''_{MS}} + 2\sum_{i=1}^{3}\frac{\alpha_i\tau_i}{1+(\omega\tau_i)^2}\right)$$

$$\tau_1^{-1} = +Q + \sqrt{a}; \qquad \tau_2^{-1} = +Q - \sqrt{a}; \qquad \tau_3^{-1} = k_R C_M + k_D,$$

$$\alpha_1 = (1/2\sqrt{a})\left\{\frac{1}{2}(k'_{MS}\,\overline{N'_{MS}} + k''_{MS}\,\overline{N''_{MS}})\right.$$

$$\times\,(k'_{MS} + k''_{MS})(P - \sqrt{a}) - \frac{J^2}{z^2e_0^2N_0}\left[\frac{1}{2}\,\frac{(k_R C_M + k_D)}{k_R C_M k_D k_S}\right.$$

$$\left.\left.\times\,\det \underset{\sim}{A}(P - \sqrt{a}) - Q + \sqrt{a} + B\left(\frac{1}{\tau_2} - \frac{1}{\tau_3}\right)\right]\right\},$$

$$\alpha_2 = -(1/2\sqrt{a})\left\{\frac{1}{2}(k'_{MS}\,\overline{N'_{MS}} + k''_{MS}\,\overline{N''_{MS}})\right.$$

$$\times (k'_{MS} + k''_{MS})(P + \sqrt{a}) + \frac{J^2}{z^2 e_0^2 N_0} \left[\frac{1}{2} \frac{(k_R C_M + k_D)}{k_R C_M k_D k_S} \right.$$

$$\left. \times \det \underset{\sim}{A}(P + \sqrt{a}) - Q - \sqrt{a} + B\left(\frac{1}{\tau_1} - \frac{1}{\tau_3}\right) \right] \Bigg\},$$

$$\alpha_3 = -(J^2/z^2 e_0^2 N_0)B,$$

$$Q = \frac{1}{2}(k_R C_M + k_D + 2k_S + k'_{MS} + k''_{MS}),$$

$$P = \frac{1}{2}(k_R C_M - k_D + 2k_S - k'_{MS} - k''_{MS}),$$

$$\alpha = P^2 + k_R C_M k_D,$$

$$\det \underset{\sim}{A} = (k'_{MS} + k''_{MS})(k_R C_M + 2k_S) + 2k_D k_S,$$

$$B = \frac{\det \underset{\sim}{A}}{(k'_{MS}+k''_{MS})(2k_S-k_D)-2k_R C_M k_S},$$

$$J = ze_0(k'_{MS}\,\overline{N'_{MS}} - k''_{MS}\,\overline{N''_{MS}}). \qquad \text{(C.5.14)}$$

The application of (C.5.14) to the experiments on nonequilibrium carrier current noise yielded agreement with the theory. Especially the difference (excess noise part) between the spectral density and the real part of the admittance came out very clearly and thus confirmed the invalidity of the fluctuation-dissipation theorem (Nyquist theorem) at nonequilibrium. We confer to the short discussion in the context of section 7.

6. Transport Fluctuations in More Sophisticated Channel Models

In the pore transport model which has been discussed in the preceding section the relevant variables in the linear phenomenological equations have been ionic occupation numbers (concentrations) at binding sites. Because interactions have been neglected, this type of variables has led to a satisfactory description of the process. In this section we shall have to go over from this macroscopic description in terms of concentrations in jump diffusion equations to a more microscopic view in terms of state variables of master equations for the single channel similarly as in the described carrier model. Now the relevant variables in the master equation are the different ionic occupation and (in section 6.3) conformational states of the channels.

The state diagrams have closed cycles, thus at nonequilibrium states admitting complex eigenvalues of the matrix of coefficients. Therefore the transport can exhibit oscillatory behaviour as consequence of the ionic interactions in the channels (Frehland and Stephan, 1979). If additionally the channels may be in different conformational states (e.g. open and closed), this can be taken into account within the discrete concept by increasing the number of states of the channels (Frehland (1979), Läuger, Stephan and Frehland (1980)).

6.1 Single-File Diffusion Through Narrow Channels

More than 25 years ago the so-called single-file mechanism for ion transport through narrow channels in biological membranes has been proposed by Hodgkin and Keynes (1955) to explain non-linear steady state current voltage characteristics of the potassium channel. Since that time a number of theoretical investigations of stationary single-file transport has been published (e.g. Heckmann (1965), (1972), Rickert (1964), Hille and Schwarz (1978). A fundamental theoretical investigation of the single-file concept concerning the time-dependent oscillatory behaviour as a function of the structural properties of the system and concerning the general structure of single-file diffusion in the limit to the continuum diffusion case has very recently been done by Stephan (1981). The essential property of single-file diffusion is that the movement of the ions is constrained to one dimension and the ions cannot overtake each other. In the usual way of mathematically describing the single-file movement the ion channel is considered as in the noninteraction case above as a sequence of binding sites separated by energy barriers over which the ions have to jump. But now the relevant equation describing the transport is the master equation for the single channel if interactions between the (identical) channels can be neglected. The single-file mechanism includes interactions between the transported ions in so far as each binding site can be occupied by only one particle and the rate constants for jumps of ions over the energy barriers are dependent on the occupation state of the pore.

As described in the noninteraction case above the transport system consists of a membrane separating two ionic solutions and containing a number N_P of identical narrow channels (pores). The concentrations of the ion species in both solutions are assumed to be helt constant. We briefly describe the main structure of the single channel master equation which determines the time dependent single-file transport: The individual channel or pore can be found in a certain number of different states depending whether the binding sites are occupied by an ion of a certain species or not. If the pore contains n binding sites and may take up p different ion species this number of states is

$$N = (1 + p)^n \tag{C.6.1}$$

These N different states are numbered by i = 1,2,...,N. With $P_\mu(t)$ denoting the time dependent probability of a single channel to be in state μ, the single channel master equation is

$$\frac{d\,P_\mu}{dt} = \sum_\nu M_{\mu\nu} P_\nu , \quad \sum_\nu P_\nu = 1 \tag{C.6.2a}$$

The matrix elements $M_{\mu\nu}$ of $\underset{\sim}{M}$ contain the transition probabilities (rate constants) for transitions from state ν to state μ. If the transport system contains many identical pores, the time dependence of the expected occupation numbers $\langle N_\mu \rangle$ of state μ, i.e. number of pores in state μ, is given by the following homogeneous set of linear first order differential equations:

$$\frac{d \langle N_\mu \rangle}{dt} = \sum_\nu M_{\mu\nu} \langle N_\nu \rangle , \quad \sum_\nu \langle N_\nu \rangle = N_P \tag{C.6.2b}$$

The structure of $\underset{\sim}{M}$ in (C.6.2) is determined by certain rules for the movement of ions, e.g. only jumps of ions between adjacent binding sites are possible, and at one instant only one jump can occur. This is expressed by the structure of state diagrams belonging to (C.6.2) (c.f. Fig. C8 and the following figures). The treatment of current noise generated by single-file systems is done within the concept of discrete transport exactly as described in section 4.

Two Binding Sites, one Ion Species

We apply the approach to the most simple example of pores with two binding sites and one ion species. In this case an individual pore can be found in four different states: As shown in Fig. C8 the empty pore is characterized by $\boxed{00}$, the pore with one ion at the first binding site by $\boxed{10}$ etc..The transition $\boxed{00} \rightarrow \boxed{10}$ is determined by the rate constant k_{00}^{10}, the number of pores in state $\boxed{00}$ is N_{00} etc..Then the time dependence for the expectation values is given by the following homogeneous set of equations:

$$\frac{d}{dt}\langle N_{00}\rangle = -(k_{00}^{10} + k_{00}^{01})\langle N_{00}\rangle + k_{10}^{00}\langle N_{10}\rangle + k_{01}^{00}\langle N_{01}\rangle$$

$$\frac{d}{dt}\langle N_{01}\rangle = -(k_{01}^{00}+k_{01}^{10}+k_{01}^{11})\langle N_{01}\rangle + k_{00}^{01}\langle N_{00}\rangle + k_{10}^{01}\langle N_{10}\rangle + k_{11}^{01}\langle N_{11}\rangle$$

$$\frac{d}{dt}\langle N_{10}\rangle = -(k_{10}^{00}+k_{10}^{01}+k_{10}^{11})\langle N_{10}\rangle + k_{00}^{10}\langle N_{00}\rangle + k_{01}^{10}\langle N_{01}\rangle + k_{11}^{10}\langle N_{11}\rangle$$

$$\frac{d}{dt}\langle N_{11}\rangle = -(k_{11}^{01}+k_{11}^{10})\langle N_{11}\rangle + k_{01}^{11}\langle N_{01}\rangle + k_{10}^{11}\langle N_{10}\rangle \qquad (C.6.3)$$

with $N_{00} + N_{10} + N_{01} + N_{11} = N_P$

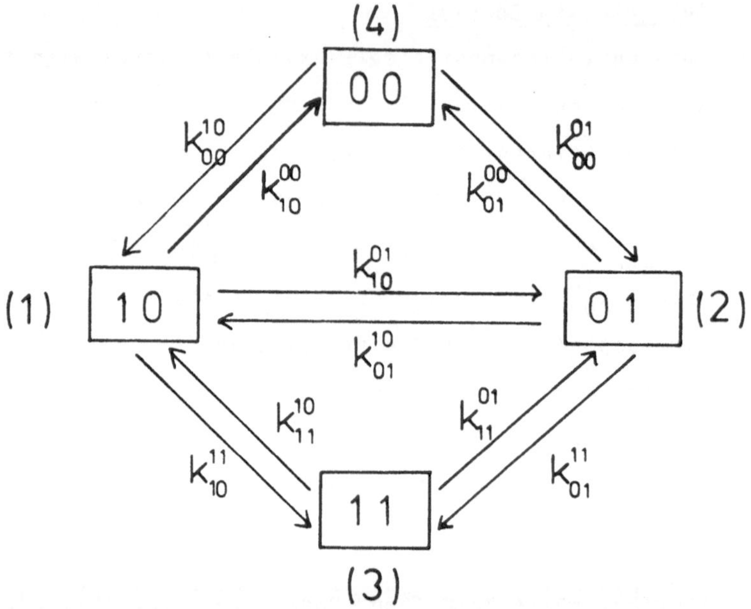

Fig. C8: State diagram for single-file diffusion in pores with two binding sites and one ion species

The four states can be numbered in the following way (see Fig. C8)

$$\boxed{10} \leftrightarrow 1, \quad \boxed{01} \leftrightarrow 2, \quad \boxed{11} \leftrightarrow 3, \quad \boxed{00} \leftrightarrow 4 \tag{C.6.4}$$

$\underset{\sim}{M}$ has the form

$$\underset{\sim}{M} = \begin{pmatrix} -(k^{00}_{10}+k^{01}_{10}+k^{11}_{10}) & k^{10}_{01} & k^{10}_{11} & k^{10}_{00} \\ k^{01}_{10} & -(k^{00}_{01}+k^{10}_{01}+k^{11}_{01}) & k^{01}_{11} & k^{01}_{00} \\ k^{11}_{10} & k^{11}_{01} & -(k^{01}_{11}+k^{10}_{11}) & 0 \\ k^{00}_{10} & k^{00}_{01} & 0 & -(k^{10}_{00}+k^{01}_{00}) \end{pmatrix} \tag{C.6.5}$$

The one vanishing eigenvalue of $\underset{\sim}{M}$ corresponds to the steady state solution (c.f. section B.3.2). For a better understanding it is useful to compare with the transport model of the preceding section for pores with two binding sites and negligible interactions between the ions: Obviously interactions may be neglected, if at all times the number of empty pores is great compared with the number of pores occupied by one or two ions. In this case the phenomelogical transport equations for the expectation values of the total numbers n_1, n_2 of ions at the first and second binding site are (c.f. (C.5.1) and Fig. C1):

$$\frac{d}{dt}\langle n_1\rangle = k_0' \cdot n_0 - (k_1'+k_1'')\langle n_1\rangle + k_2''\langle n_2\rangle$$

$$\frac{d}{dt}\langle n_2\rangle = k_1'\langle n_1\rangle - (k_2'+k_2'')\langle n_2\rangle + k''_3 n_3 \tag{C.6.6}$$

Clearly, one gets approximately the jump diffusion equations (C.6.6) from (C.6.3) in the limit

$$k_{00}^{10} << k_{10}^{00}, \quad k_{00}^{01} << k_{01}^{00},$$

$$k_{10}^{11} << k_{11}^{10}, \quad k_{01}^{11} << k_{11}^{01} \tag{C.6.7}$$

As consequence:

$$N_{10} \simeq n_1 , \; N_{01} \simeq n_2 , \; N_{11} \simeq 0. \tag{C.6.8}$$

Another important limiting case is the case where the state $\boxed{11}$ is forbidden (one ion case). It has been proposed (Frehland and Läuger (1974)) that possibly in some cases for electrostatic reasons a single pore can be occupied by only one ion. Clearly, in this case N_{10} is equal to the number of ions at binding site 1 and N_{01} at binding site 2:

$$N_{10} = n_1 , \; N_{01} = n_2. \tag{C.6.9}$$

In this case one can get two transport equations for n_1 and n_2 by setting $N_{11}=0$ and replacing N_{00} by $(N_P-n_1-n_2)$

$$\frac{d}{dt}\langle n_1\rangle = k_1\left(1-\frac{\langle n_1\rangle+\langle n_2\rangle}{N_P}\right)n_1 - (k_1'+k_1'')\langle n_1\rangle + k_2''\langle n_2\rangle$$

$$\frac{d}{dt}\langle n_2\rangle = k_r\left(1-\frac{\langle n_1\rangle+\langle n_2\rangle}{N_P}\right)n_r - (k_2'+k_2'')\langle n_2\rangle + k_1'\langle n_1\rangle \quad \text{(C.6.1}$$

A general discussion of these limiting cases for arbitrary numbers of binding sites hat been given by Stephan (1981). In the following numerical calculations of spectra for special examples we shall include both limiting cases.

6.2 Numerical Calculations of Single-File Noise in Open Channels

We present some explicit calculations of current noise spectral densities generated by single-file ion transport in pores with two binding sites by evaluating (C.4.5). The explicit way of calculati of $\underset{\sim}{\Omega}$ and of spectral density $G_{\Delta J}$ including the case of complex eigenvalues has been described (Frehland and Stephan, 1979).

As in the noninteraction case in section 5 the dependence of the spectral density on different parameters as the transition probabilities or the constants $\gamma_{\mu\nu}$ determining the contributions of the individual transitions to the measured current is rather complicated. Again some typical dependences and properties are worked out.

Equilibrium

Calculations of spectra of current noise at equilibrium ($J^s=0$) are presented in Fig. C9.Case a) and c) simulate the two limiting cases of very strong electrostatic interactions between the ions and negligible interactions. In the latter example where the probabilities for entrance of ions into the pores are small, we take into account this fact by multiplying $G_{\Delta J}$ with a factor 100, assuming that the number of pores in this case is a hundred times greater as in the other cases. Hence the results for $G_{\Delta J}$ in C9,though in arbitrary units, are normalized to the condition that the average number of ions within all pores is approximately equal. The shape of the three spectra shows no essential differences.

At equilibrium the rate constants for closed loops satisfy the so-called Wegscheider conditions (Bak (1959), Higgins (1967), Hearon (1953)) which are equivalent to the principle of microscopic reversibility (Bak, 1959). In this case all eigenvalues of $\underset{\sim}{M}$ are real and nonpositive. Hence at equilibrium the single-file transport is nonoscillatory and the spectra show no peaks.

Nonequilibrium

At non-equilibrium the eigenvalues of $\underset{\sim}{M}$ may be complex because the Wegscheider-conditions are no longer satisfied and the state diagrams represent driven cycling steady states. This means that the single-file transport possibly may exhibit damped oscillatory behaviour. As consequence the resulting spectra may show peaks. As has been suspected since many years and recently could be proven

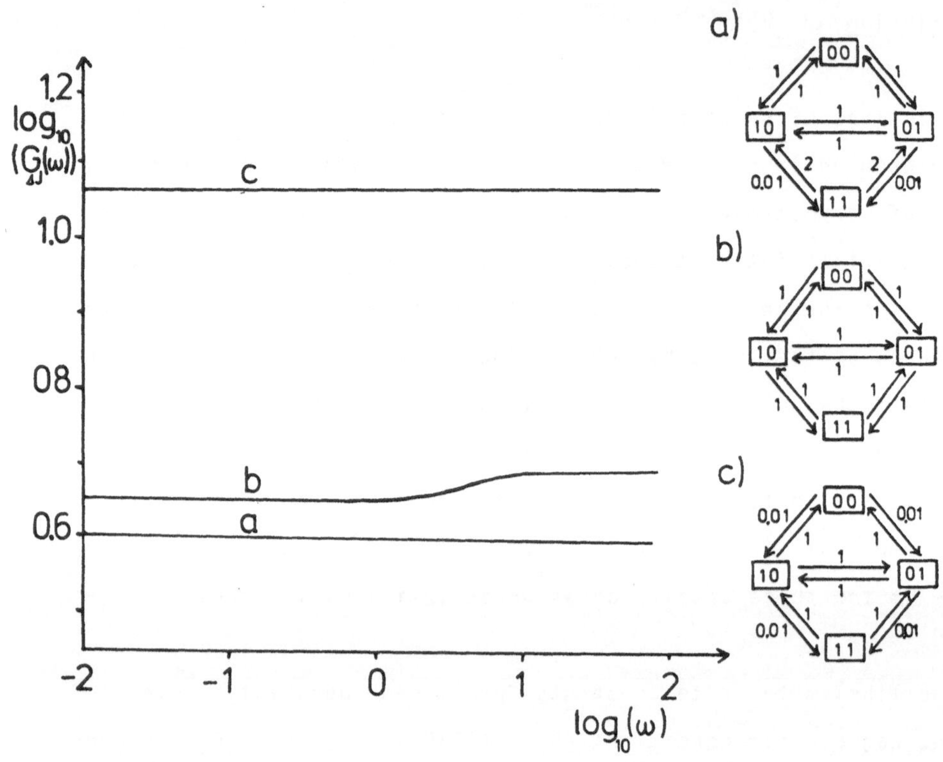

Fig. C9: Spectra $G_{\Delta J}$ of current noise at equilibrium ($\overset{s}{J}=0$) for different transport models (arbitrary units). The state diagrams illustrate the different model situations being calculated a) one ion case, b) general single-file model with equal transition probabilities between all four states, c) case of usual jump diffusion, because the probability of pores to be empty (state 00) is great compared with the probability of pores to be occupied. The contributions of the individual transitions to the measured current, which are determined by the constants $\gamma_{\mu\nu}$, are set equal.

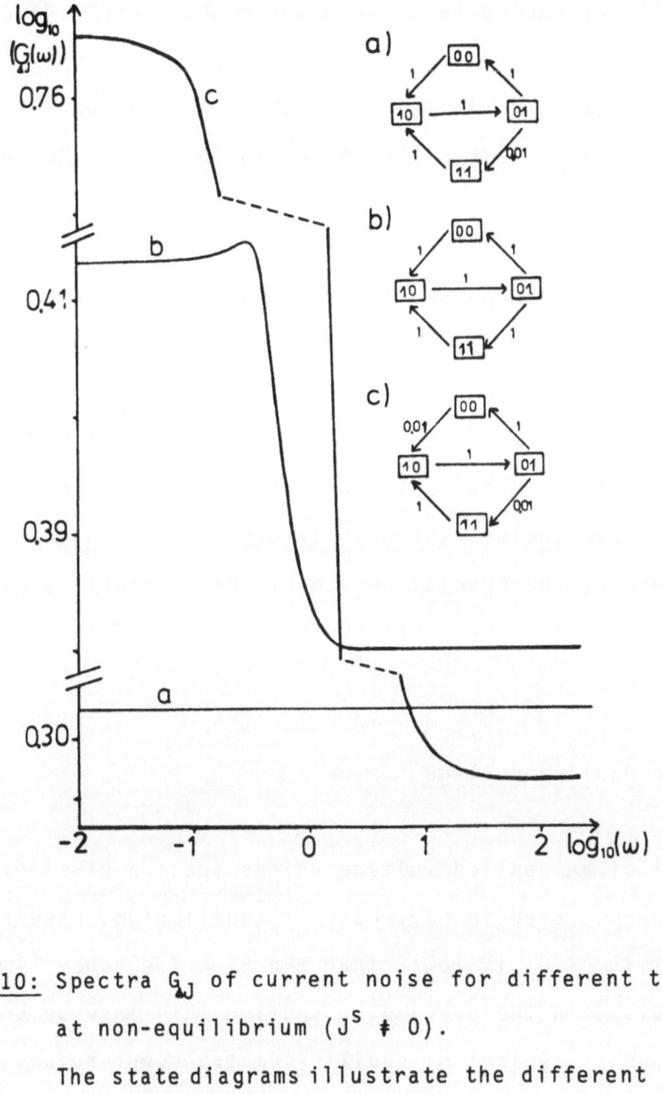

Fig. C10: Spectra $G_{\Delta J}$ of current noise for different transport models at non-equilibrium ($J^S \neq 0$).
The state diagrams illustrate the different calculated model situations. The applied voltage is assumed to be high, so that jumps of ions take place only in one direction.

by Stephan (1981) for one-cycle state diagrams the weakest damping of oscillations occurs in those cases where transitions only in one direction are possible. In this case the effect of oscillatory transport is expected to give the greatest influence on the course of the spectra.

Some typical features of non-equilibrium single-file noise are demonstrated by the numerical results in Fig. C10. In case b) the spectrum shows a peak as consequence of the oscillatory behaviour. In case a) the spectrum is white, though two eigenvalues of $\underset{\sim}{M}$ are complex and hence the fundamental solution matrix $\underset{\sim}{\Omega}$ has an oscillatory time dependence. Nevertheless the oscillatory terms in $G_{\Delta J}$ cancel each other because of the special choice of the constants $\gamma_{\mu\nu}$. In the case c) of negligible interactions the spectrum exhibits a Lorentz-type behaviour.

The Effect of an Applied Voltage

A typical effect of an applied voltage on the spectra of single-file noise is demonstrated in Fig. C11. At equilibrium (case a) the low frequency limit of $G_{\Delta J}$ is lower than the high frequency limit. As already discussed in the preceding section this inverse Lorentz-spectrum behaviour is typical of equilibrium transport noise.

The state diagram in Fig.C11 shows a special choice of rate constants at equilibrium (case a)).With the ansatz for the rate constants (c.f. (C.5.11))

$$k_{\mu\nu} = \bar{k}_{\mu\nu}\, e^{\gamma_{\mu\nu} \cdot u} \qquad (C.6.11)$$

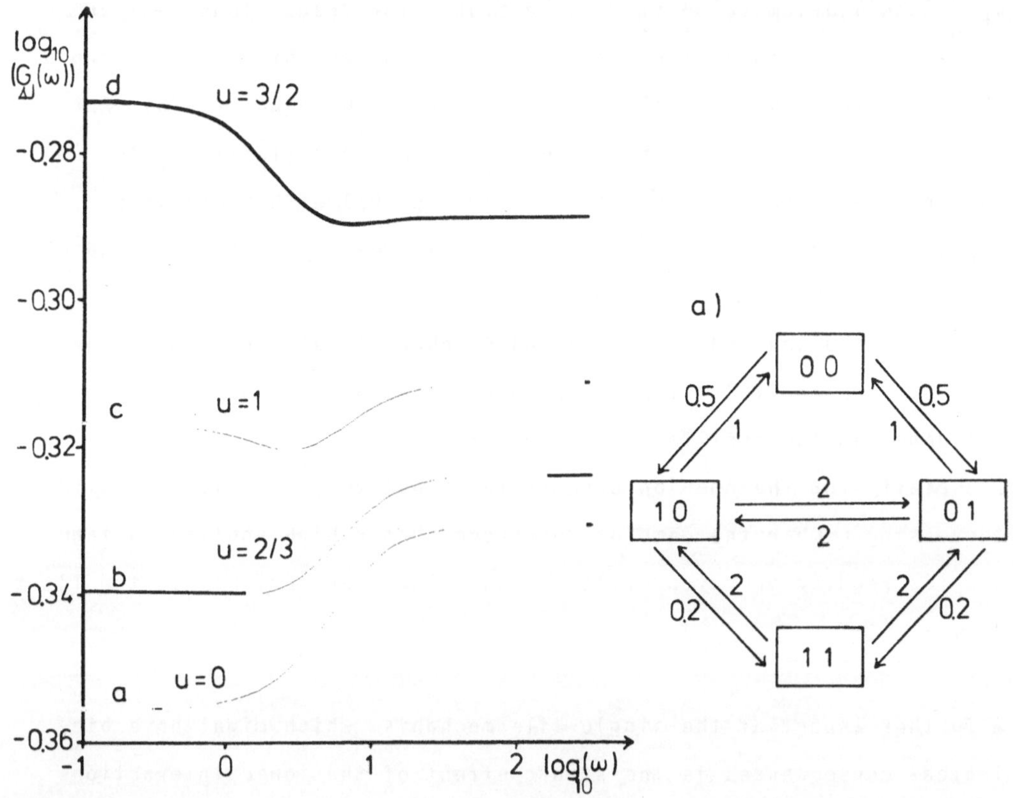

<u>Fig. C11:</u> The effect of an applied voltage on the spectrum $G_{\Delta J}$ of current noise generated by a single-file transport model. The change of jump constants under the influence of different applied voltages is taken into account by different values of u according to (C.6.11).

($\bar{k}_{\mu\nu}$: equilibrium value of rate constants for transitions $\mu\to\nu$) an increasing applied voltage can be taken into account by increasing u. The absolute values of all $\gamma_{\mu\nu}$ are assumed to be equal (=1/3). For transitions connected with ionic jumps to the right (left) $\gamma_{\mu\nu}$ is chosen positive (negative). As shown by Fig. C11 with increasing applied voltage the low frequency limit increases. In case b) and c) the spectra show (weak) minima.

As has been discussed (Frehland and Stephan, 1979) and in more detail by Stephan (1981) the number n of binding sites has a strong influence on the oscillatory behaviour.

Especially for the one ion case it turns out very clearly that with increasing number the damping decreases. Under high applied voltage and for regular energy profiles the damping even vanishes in the limit $n\to\infty$.

A further aspect of the single-file mechanism which might have biological consequences is the strong effect of the ionic interactions on the absolute intensity of current noise. This will be discussed in some detail in section 8.

6.3 Single-File Noise: Coupling Between Transport and Open-Closed Kinetics of the Channels

We now will discuss the problem of transport noise generated by ionic channels which additionally may assume different conducting states (open and closed).

In the limit of fast ionic movement compared with the open-closed kinetics of the channels we should get agreement with the usual channel noise treatment (c.f. section 3.7 in part B) from the master equation approach.

State diagrams and kinetic equations

First we want to make some remarks how to arrive at equations describing the ion transport through channels with open-closed kinetics. As above the single channel in a special state is considered as a sequence of binding sites. The individual channel can be in a certain number of different states which generally depends a) on the number of ion species which may penetrate the channel, b) on the number of binding sites within the channel and c) on the special structure of channel kinetics, e.g. the number of different conductance states. In order to illustrate how to get state diagrams for special models we have shown in Fig. C 12 the state diagram for transport of one ion species through channels with two binding sites and two-state (open-closed) channel kinetics. The open channel may be in four states

[op | 00] , [op | 10] , [op | 01] and [op | 11] , where [op | 00] is the unoccupied open channel, [op | 10] the open channel with an occupied first binding site etc.. Correspondingly, the closed channel may be in four states. The interior part of the state diagram in Fig.C 12 is the usual single-file diagram for pores with two binding

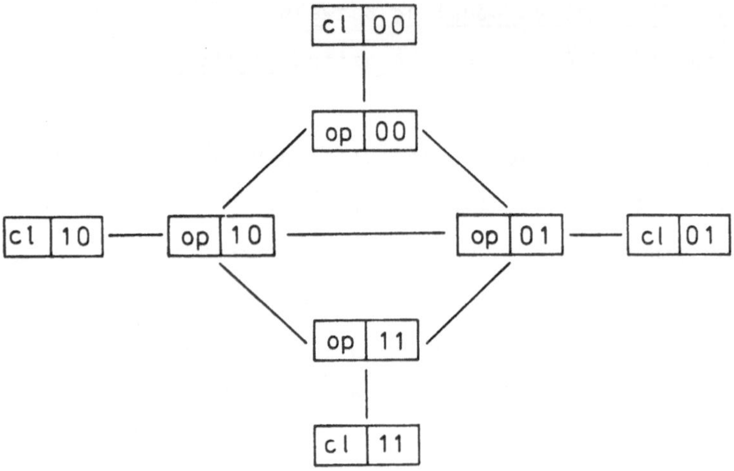

Fig. C12: State diagram for channels with two-state open-closed kinetics, two binding sites and one ion species.

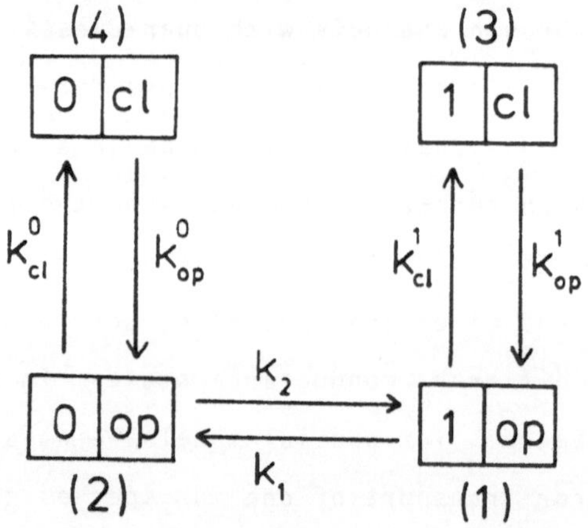

Fig. C13: State diagram for channels with open-closed kinetics and one binding site.

sites (cf. Fig.C8). Under the assumption, that in closed channels an ionic movement is not possible, transitions between the four different closed states are forbidden. On the other hand the open-closed kinetic of the channel may be dependent on its occupation state. Obviously this point can strongly influence the transport kinetics, a point which will shortly be discussed below in this section.

In Fig. C13 is shown the corresponding state diagram for the most simple case of pores with one binding site only.

The single channel may be in four different states:

(1) ,(occupied, open) , [1 | op]
(2) ,(unoccupied, open), [0 | op]
(3) ,(occupied, closed) , [1 | cl]
(4) ,(unoccupied, closed) , [0 | cl] (C.6.11)

We distinguish between rate constants k_{cl}^{o}, k_{op}^{o} and k_{cl}^{1}, k_{op}^{1} for open-closed transitions according to the occupation state of the channel. Transitions between the open-unoccupied and open-occupied states are determined by jumps of ions from and to the left, right sides of the channel, respectively.
Therefore according to Fig.C13

$$k_1 = k' + k''$$
$$k_2 = k_l' + k_r'' \qquad (C.6.12)$$

If P_μ denotes the probability of a channel to be in state μ , the single-channel master equation is

$$\frac{dP_1}{dt} = -(k_1+k_{cl}^1)P_1 + k_2N_2 + k_{op}^1P_3$$

$$\frac{dP_2}{dt} = -(k_2+k_{cl}^o)P_2 + k_1P_1 + k_{op}^oP_4$$

$$\frac{dP_3}{dt} = -k_{op}^1P_3 + k_{cl}^1P_1$$

$$\frac{dP_4}{dt} = -k_{op}^oP_4 + k_{cl}^oP_2 \qquad (C.6.13)$$

N_p is the total number of channels, which is assumed to be constant.

If the ionic movement through the channels is fast compared with the open-closed kinetics, i.e.

$$k_1,\ k_2 \gg k_{op}^o,\ k_{op}^1,\ k_{cl}^o,\ k_{cl}^1 \ , \qquad (C.6.14)$$

(C.6.13) becomes approximately

$$\frac{dP_{op}}{dt} = -\left(\frac{k_2}{k_1+k_2}k_{cl}^1 + \frac{k_1}{k_1+k_2}k_{cl}^o\right)P_{op}$$

$$+ k_{op}^1P_3 + k_{op}^oP_4 \ ,$$

$$\frac{dP_3}{dt} = - k_{op}^1 P_3 + k_{cl}^1 \frac{k_2}{k_1+k_2} P_{op} \quad ,$$

$$\frac{dP_4}{dt} = - k_{op}^o P_4 + k_{cl}^o \frac{k_1}{k_1+k_2} P_{op} \quad , \qquad \text{(C.6.15)}$$

with

$$P_{op} = P_1 + P_2, \qquad P_1 = \frac{k_2}{k_1+k2} P_{op} = \frac{k_2}{k_1} P_2 \quad ,$$

$$P_{op} + P_3 + P_4 = 1 \qquad \text{(C.6.16)}$$

For channels with more than one binding site the more complex kinetic equations can be derived in correspondence to (C.6.13). For fast ionic movement the equations are simplified analogously as (C.6.15).

We should make a short remark, concerning the structure of γ. E.g. we can assume that the opening and closing of the channel (in Fig. C13 these are the transitions $1 \rightarrow 3$ and $2 \rightarrow 4$) does not yield a contribution to the electric current. Though below we shall make this assumption for simplicity, we emphasize that it is not necessary. Movement of charges connected with the opening and closing of the channels, i.e. so-called gating currents and the corresponding contribution to the electrical fluctuations can be included without principal complication.

Though the one-binding site model is most simple, just in this case a complication arrises, because transitions between states 1 and 2 are not uniquely connected with one special type of jumps of particles. As may be seen from Fig.C13, transitions $1 \rightarrow 2$ from the occupied open channel to the unoccupied open channel can be generated by jumps of ions to the left (k") as well as to the right (k') and transitions $2 \rightarrow 1$ correspondingly. We shall give the explicit formula for this special case below. In the general treatment we shall exclude this case and assume that transitions $\mu \rightarrow \nu$ are connected with only one type of ionic jumps and hence a unique contribution to the electric current. The reader who wants to treat more sophisticated models can do it by increasing the number of individual fluxes $\phi_{\mu\nu}$ and applying the formalism correspondingly.

Simplified equations for fast transport

We now make the restricting assumption that the open-closed kinetics are not connected with (gating) currents. As mentioned, for channels with one conducting (open) state, subdivided only by the different ionic occupation states, the equations (C.6.13) can generally be simplified by introduction of the total number N_{op} of open channels, if the ionic movement across the open channels (transport kinetics) is fast compared with the channel open-closed kinetics. For one channel the probabilities for the different ionic occupation states of the open channel are given by the probability of the channel to be open.

Hence, in the equations, which are simplified under the assumption of fast ionic movement, occurs one probability P_{op} characterizing the open channel while the other variables characterize closed states of the channel which are distincted by the different ionic occupation states. Generally we write these approximate equations in the form

$$\frac{dP^*_\mu}{dt} = \sum_{\nu=1}^{M} M_{\mu\nu} P^*_\nu , \quad M<N \qquad (C.6.17)$$

with $P^*_1=P_{op}$. They describe the channel kinetics with the time-dependent fast transport kinetics being neglected.

Analogously we can introduce the fundamental solution matrix $\underset{\sim}{\Omega}^*(t)$ of (C.6.17).

Thus

$$\overline{\Omega^*_{11}} = \Omega_{op} + P_1^{S*} , \quad \Omega_{op} := \Omega^*_{11} \qquad (C.6.18)$$

is the probability for the single channel to be open under the initial condition to be open at t=0. As we shall see, Ω_{op} essentially determines the electrical fluctuations.

<u>Current through the single channel</u>

The assumption, that the transport kinetics are fast, implies the assumption that the expectation value $<j_s>$ of electric current j_s through the single open channel may be treated as if stationary. Obviously $<j_s>$ can be expressed by the stationary current J^s divided by the steady state number N^s_{op} of open channels. Then $<j_s>$ is given by

$$<j_s> = \frac{J^s}{N^s_{op}} = \sum_{\mu,\nu} \gamma_{\mu\nu} \; M_{\mu\nu} \; \hat{\Omega}^s_\nu , \; \hat{\Omega}^s_\nu = \frac{N^s_\nu}{N^s_{op}} . \qquad (C.6.19)$$

If ν is an ionic occupation state of the open channel, $\hat{\Omega}^s_\nu$ is the expectation value of this state for the open channel, which is approximately stationary.

For sufficiently small applied voltages V, where $<j_s>$ is proportional to the applied voltage, we can introduce the single channel conductance Λ :

$$<j_s> = \Lambda \cdot V \qquad (C.6.20)$$

Current fluctuations

The current fluctuations contain a shot noise contribution and a term which is time-, frequency-dependent respectively. It is this term which we expect to agree with the results from the usual master equation approach to channel fluctuations under the condition of fast ionic movement. Under this condition the fundamental solutions $\Omega_{\rho\mu}(t)$ in (C.4.5.) consist of terms with very fast time constants generated by the transport kinetics and terms with time constants coming from the channel kinetics. If we want to apply the approximate solution $\underset{\sim}{\Omega}^*$, the terms containing the fast transport kinetics yield an additional white noise term which must be added to the first sum in (C.4.5),(C.4.6). We do not further consider this term. Under the simplifying assumption that only from jumps of the ions over the barriers of the open channel we get contributions to the electric current and hence nonvanishing $\gamma_{\mu\nu}$, $\gamma_{\kappa\rho}$, the remaining components $\Omega_{\rho\mu}(t)$ belong to open channel states, which approximately are given by Ω_{op} through

$$\Omega_{\rho\mu}(t)= \hat{\Omega}^{s}_{\rho} \cdot \Omega_{op}(t) \tag{C.6.21}$$

and independent of μ. Therefore, with (C.6.19) we get the approximation

$$\sum_{\mu,\nu,\kappa,\rho} \gamma_{\mu\nu}\gamma_{\kappa\rho}\, \phi^{S}_{\mu\nu}\, M_{\kappa\rho}\,\Omega_{\rho\mu}(t) = N^{S}_{op}\langle j^{2}_{s}\rangle\,\Omega_{op}(t)+A_{1}\delta(t) \tag{C.6.22}$$

A remark, important for the discussion in section 7, should be made: (C.6.21) means that because of the fast ionic movement $\Omega_{\rho\mu}(t)$ becomes independent of μ. Hence $\gamma_{\mu\nu}\, M_{\mu\nu}\, N^{S}_{\nu}$ in (C.6.22) uncouples from $\Omega_{\rho\mu}$ and the summations over μ,ν and κ,ρ can be made independently. Because $\underset{\sim}{\gamma}$ is antisymmetric , generally holds:

$$J^{S} = \sum_{\mu,\nu}\gamma_{\mu\nu}\,\phi^{S}_{\mu\nu} = \sum_{\mu,\nu}\gamma_{\mu\nu}\,\frac{1}{2}(\phi^{S}_{\mu\nu}-\phi^{S}_{\nu\mu}) \tag{C.6.23}$$

J^{S} is determined by the antisymmetric part of the stationary flux matrix $\underset{\sim}{\phi}^{S}$. Therefore the sum in (C.6.22) and the time-

(frequency-) dependent part of the current fluctuations in (C.6.24), (C.6.25) is also given only by the antisymmetric part of $\underset{\sim}{\phi}^{s}$. Hence the channel-noise in the fast-movement case is according to (C.4.6.) given by the excess-noise nonequilibrium term.

Autocorrelation function and spectral density of current fluctuations are with (C.4.5), (C.4.6), (C.6.22).

$$<\Delta J(o)\,\Delta J(t)> = A\cdot\delta(t) + N^{s}_{op}<j_s>^2\,\Omega_{op}(t) \qquad (C.6.24)$$

and

$$G_{\Delta J}(\omega) = 2A + 4\,N^{s}_{op}<j_s>^2\cdot\int_0^{\infty}\Omega_{op}(t)\cos\omega t\,dt \qquad (C.6.25)$$

The white noise term A is composed of the first sums and the contributions from the fast terms in $\Omega_{\rho\mu}(t)$ in (C.4.5), (C.4.6) respectively. If (C.6.20) is valid, $<j_s>$ can be expressed by the single channel conductance Λ and applied voltage V. (C.6.24),(C.6.25) express the general result , that in the equilibrium ($<j^{s}>=0$) the current fluctuation exhibit freqency independent white noise only. This is a consequence of the condition of fast ionic movement. Below in the numerical examples we shall see explicitely that the spectra may become frequency-dependent, if we leave this condition.

A comparison of the time-(frequency-)dependent "channel - noise" terms with the result (B.3.41) from the master equation approach immediately shows agreement, if we regard that

$$\Omega_{op}^{(t)} = \Omega_{11}^{(t)}$$

and

$$N_{op}^{s} = P_1^{s} \cdot N_p$$

The agreement between the two rather different approaches in the 'overlapping' region of fast transport may serve a successful test of consistency.

Naturally, both approaches may easily be extended to channel models with more than one conducting (open) state and agree also under these more complicated conditions (Frehland, 1979).

Channels with one Binding Site

As a special example without restriction to fast ionic movement we discuss the case of channels with one binding site and simple open-closed kinetics shown in Fig.C13 and described by the kinetic equations (C.6.13). As mentioned above for transitions $1 \rightarrow 2$ we must distinguish between jumps to the left and to the right and for transitions $2 \rightarrow 1$ correspondingly. Taking into account this fact and considering the correlations between the individual fluxes one gets an autocorrelation function

$<\Delta J(o)\,\Delta J(t)> =$

$$\left[\gamma_1^2(k''N_1^s+k_l'N_2^s) + \gamma^2(k'N_1^s+k_r''N_2^s)\right]\delta(t)+$$

$$+ \left[(\gamma_1 k'' - \gamma_2 k')(\gamma_2 k_r'' - \gamma_1 k_l')(N_1^s \Omega_{22} + N_2^s \Omega_{11})\right.$$

$$\left. + (\gamma_2 k_r'' - \gamma_1 k_l')^2 N_2^s \Omega_{21} + (\gamma_1 k'' - \gamma_2 k')^2 N_1^s \Omega_{12}\right] \qquad (C.6.26)$$

γ_1 is the contribution to the electric current for jumps over the first barrier and γ_2 over the second barrier, positive sign for jumps to the right, negative sign for jumps to the left. The spectral density is

$$G_{\Delta J} = 2\left[\gamma_1^2\,(k''N_1^s+k_l'N_2^s)+\gamma_2^2(k'N_1^s+k_r''N_2^s)\right] +$$

$$+4\int_0^\infty dt\,\cos\omega t\,\left[(\gamma_1 k'' - \gamma_2 k')(\gamma_2 k_r'' - \gamma_1 k_l')(N_1^s\Omega_{22}+N_2^s\Omega_{11})\right.$$

$$\left.+(\gamma_2 k_r'' - \gamma_1 k_l')^2 N_2^s \Omega_{21} + (\gamma_1 k'' - \gamma_2 k')N_1^s\Omega_{12}\right] . \qquad (C.6.27)$$

The fluctuations around equilibrium not neccessarily yield white noise, if the characteristic times of ionic transport are comparable to the times of open-closed kinetics.

In order to illustrate the effect of coupling between transport kinetics and channel open-closed kinetics on the current fluctuations, we give some numerical results for the spectral density, using (C.6.27. We set $k_1 = k_2 = 1$,

$$k^o_{op} = k^1_{op} := k_{op},\ k^o_{cl} := k_{cl}, \qquad (C.6.28)$$

This is a situation where the channel open-closed kinetics are independent of the ionic occupation state of the channel. The ratio between the characteristic times of channel kinetics and of transport kinetics is varied in Figures C14 and C15 through variation of $k_{op} = k_{cl}$ from fast open-closed kinetics (curves a) to slow open-closed kinetics (curves d). Figure C14 shows results for fluctuations at equilibrium ($J^S=0$) and Figure C15 at non-equlibrium ($J^S \neq 0$). In the equilibrium case we have set $k^{'}=k^{''}$ and $k^{'}_{l} = k^{''}_{r}$, at nonequlibrium $k_1 = k^{'}$. $k_2 = k^{'}_{l}$. Furthermore, in the equilibrium case we have chosen $\gamma_1=\gamma_2$. This means that the channel has asymmetric properties and hence according to (C.6.27) $G_{\Delta J}$ is frequency-dependent.

If the channel kinetics are fast compared with the transport kinetics (cases a) in Figures C14 and C15 the transition between low and high frequency limit is determined by the transport kinetics. If the channel kinetics are slow (curves d), the spectrum in the nonequilibrium case shows a Lorentz-type behaviour over a wide frequency region, generated by the open-closed kinetics. The coupling of transport kinetics and channel open-closed kinetics is shown in the curves b,c where the characteristic times are comparable. The high frequency limit in all cases is independent of the coupling. Comparing the numerical results we see, that at equilibrium the low frequency limit is lower (equal) than the high frequency limit and at nonequilibrium vice versa.

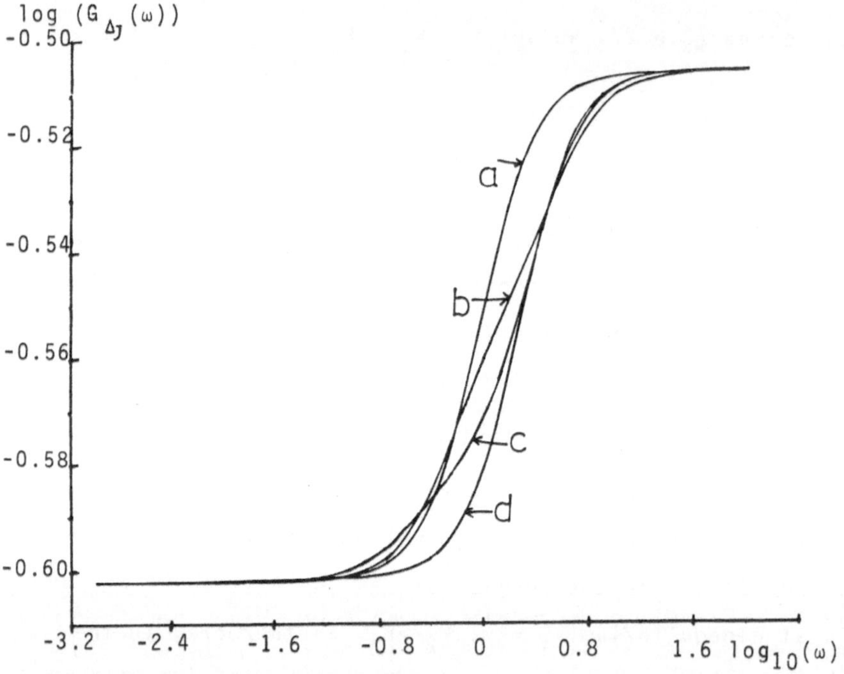

Fig. C14: Numerical results for spectral density $G_{\Delta J}$ of current fluctuations around equilibrium generated by channels with one binding site and simple open-closed kinetics. $k_1 = k_2 = 1$, $\gamma_1 = 3/4$, $\gamma_3 = 1/4$, $k' = k'' = k_l' = k_r'' = 1/2$. $k_{op} = k_{cl}$ are 100, 1, 1/3, 1/100 for curves a, b, c, d respectively.

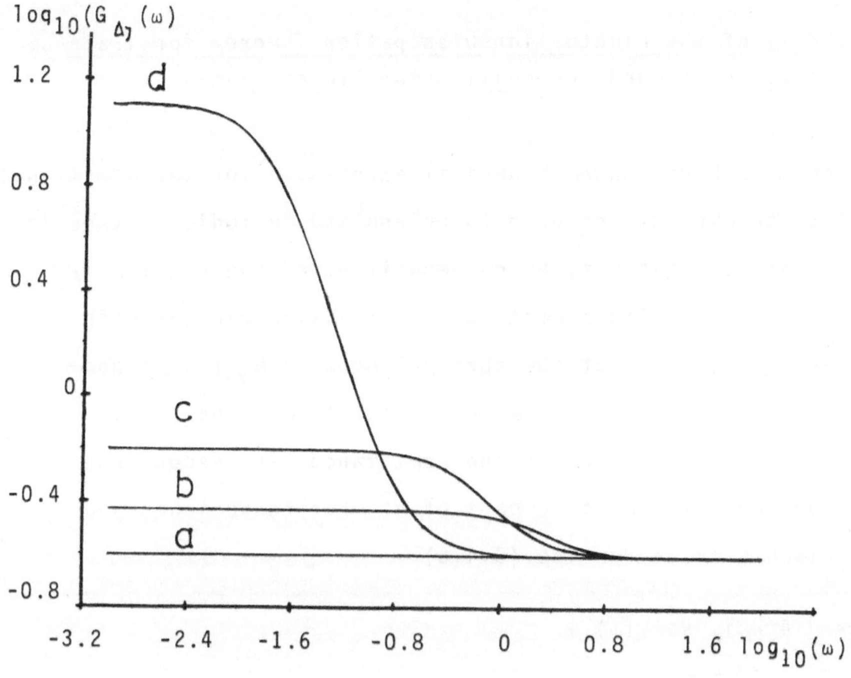

Fig. C15: Numerical results for spectral density $G_{\Delta J}$ of current fluctuations at nonequilibrium. All parameters for the four curves a,b,c, d are chosen as in Fig.C14 except $\gamma_1=\gamma_2$ and $k_1=k'$, $k_2=k'$.

7. Invalidity of the Fluctuation-Dissipation Theorem for Transport Fluctuations Around Non-equilibrium Steady States

In section 3 we have derived a general expression for the admittance describing the current response to an applied periodic voltage in discrete transport systems. By decomposition of the stationary flux matrix $\underset{\sim}{\phi}^s$ into its symmetric and antisymmetric parts in (C.4.6) we could show that the spectral density $G_{\Delta J}(\omega)$ at non-equilibrium is determined by two different terms: the first term contains the real part of the admittance, the second one is given by the antisymmetric part of $\underset{\sim}{\phi}^s$. For convenience and further discussion we rewrite (C.4.6)

$$G_{\Delta J}(\omega) = 4\, k_b T_a \operatorname{Re} y(\omega) + $$
$$+ 4 \sum_{\mu,\nu,\kappa,\rho} \gamma_{\mu\gamma}\gamma_{\kappa\rho} \frac{1}{2} (\phi^s_{\mu\nu} - \phi^s_{\nu\mu}) M_{\kappa\rho} \int_0^\infty \Omega_{\rho\mu}(t) \cos \omega t dt \qquad \text{(C.7.1)}$$

At equilibrium defined by detailed balance (C.2.8) (symmetry of $\underset{\sim}{\phi}^s$) the second term vanishes and by (C.7.1) the validity of the fluctuation-disstpation theorem (Nyquist theorem) (C.3.2) is confirmed. The result can be applied to other transport observables T with (C.2.9.) than the electric current, if a relation of the kind (C.3.8) is satisfied, which yields a strong connection between the action of the external force on the rate constants and the linear (transport) coefficients $\gamma_{\mu\nu}$. We recently have proposed to call such forces <u>conjugate</u> with respect to T (Frehland 1981). Within our framework we supposed this to be a sensible definition of the forces to be conjugate, because a relation of the type (C.3.8) is essential for the response (admittance) function to satisfy a fluctuation-dissipation relation at equilibrium. We conjecture that there is agreement with the usual definition of conjugate forces in irreversible thermodynamics.

At nonequilibrium the second term in (C.7.1) does not vanish. The discussion of the general structure of this term is facilitated by a slight transformation: The expression

$$J^s_{\mu\nu} = \gamma_{\mu\nu}(\phi^s_{\mu\nu} - \phi^s_{\nu\mu}) = J^s_{\nu\mu} \qquad \text{(C.7.2)}$$

is symmetric because $\underset{\sim}{\gamma}$ is antisymmetric. It is the stationary contribution to J^s by transitions $\nu\to\mu$ and $\mu\to\nu$. Furthermore

$$\Delta J_\mu(t) = \sum_{\kappa,\rho,\mu} \gamma_{\kappa\rho}\, M_{\kappa\rho}\, \Omega_{\rho\mu}(t) \qquad \text{(C.7.3)}$$

is the time dependent deviation of total current J from the stationary current J^s under the initial condition, which distinguishes state μ at t=o. Then we get from (C.7.1)

$$G_{\Delta J}(\omega) = 4k_bT_a\mathrm{Re}Y(\omega) + 2\sum_{\mu,\nu} J^s_{\mu\nu}\int_0^\infty \Delta J_\mu(t)\cos\omega t\,dt \qquad \text{(C.7.4)}$$

The interpretation is as follows: The first (admittance) term is generated by microscopic disturbances of the system which can also be excited macroscopically by an external voltage.

It contains no additional information compared with the macroscopic admittance function. On the other hand the second term is generated by (microscopic) fluctuations which cannot be excited by an external voltage. Because of this term the measurement of nonequilibrium current fluctuations in principle may yield information about the transport system which cannot be obtained from the measurement of macroscopic quantities.

7.1. Special Examples: Channel Noise, Carrier Noise, Noise Generated by Permanently Open Channels

We illustrate the meaning of the second nonequilibium noise term at some examples, which have already been investigated in the course of this volume.

Channel Noise

The second term in (C.7.4) is a bilinear expression in current, and we may expect that its contribution to fluctuations often increases quadratically with current (voltage). Indeed this property is typical for nonequilibrium excess noise as e.g. Lorentzian noise generated by nerve channels with open-closed kinetics, discussed in the preceding section. Here we have seen that in case the channels possess one open state and the ion transport through the channels is fast compared with the opening-closing kinetics of the channels, the current noise spectral density has the general structure

$$G_{\Delta J}(\omega) = \text{white noise} + \text{Lorentzian Noise}$$

The Lorentzian term is proportional to the square of the current through the single open channel. Comparison of the results of the preceding section with (C.3.17),(C.7.4.) shows that the white noise is determined by the admittance which in this case is frequency-independent because the opening and closing have been assumed to yield no contribution to the electric current and hence via (C.3.8) to be voltage-independent.

On the other hand, if the opening and closing of the channels is voltage-dependent,it is connected by (C.3.8) with a current contribution, the so-called gating current. In this connection the previous results lead to an interesting note, concerning the different types of information contained in the real part of the admittance and the noise spectrum: While in the noise spectrum the Lorentzian channel noise is expected to be dominant, the real part of the admittance does not contain such a contribution and its frequency dependence is expected to be generated by the gating process. Hence the measurement of the admittance (see e.g. Fishman et al., 1977, 1981, Moore et al., 1980) possibly yields information about the gating process.

Carrier Noise

The noise generated by carrier mediated ion transport has been discussed in section 5.3. The experimental results recently published (Kolb a. Frehland, 1980) have confirmed our prediction

---

*) Indeed, very recent results of Fishman et al. (1982, preprint) seem to confirm this prediction.

of the breakdown of the fluctuation dissipation theorem. An example is given at Fig. C 16

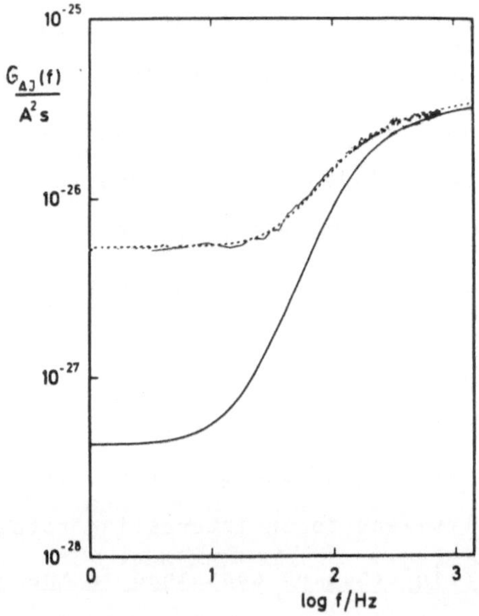

Fig. C 16: Spectral density of current noise from a lipid bilayer membrane in the presence of 10^{-7}M tetranactin at an applied voltage of 100 mV. The dotted line is a fit to the experimental curve with the use of (C.5.14). The lower curve is 4 $k_b T_a$ Re $Y(\omega)$, which would be predicted as spectral density, if the fluctuation dissipation theorem would be valid also at nonequilibrium.

Permanently open pores with one binding site

The case of permanently open pores with one binding site within the pore, one ion species and negligible interactions between the ions and pores can also serve as a simple example for discussion (c.f. section 1.2 and Fig.Co).

The spectral density of current fluctuations is given by (C.1.9), (C.1.10)

$$G_{\Delta J}(\omega) = 2A_1 + A_2 \frac{4\tau}{(1+\omega^2\tau^2)} \qquad \text{(C.7.5)}$$

with

$$A_1 = \gamma_1^2(k_o'N_o + k_o''N_1^s) + \gamma_2^2(k_1'N_1^s + k_2''N_2)$$

$$A_2 = (\gamma_1 k_o'N_o - \gamma_2 k_2''N_2)(\gamma_2 k_1' - \gamma_1 k_1'') \qquad \text{(C.7.6)}$$

For the admittance $Y(\omega)$ holds with (C.5.10)

$$4\, k_bT_a \mathrm{Re}Y(\omega) = 2A_1 + \frac{2\tau}{1+\omega^2\tau^2} (\gamma_2 k_1' - \gamma_1 k_1'') \cdot$$

$$\left\{\gamma_1(k_o'N_o + k_1''N_1^s) - \gamma_2(k_2''N_2 + k_1'N_1^s)\right\} \qquad \text{(C.7.7)}$$

The decomposition of $G_{\Delta J}(\omega)$ is

$$G_{\Delta J}(\omega) = 4\, k_bT_a \mathrm{Re}Y(\omega) + J^s(\gamma_2 k_1' - \gamma_1 k_1'') \frac{2\tau}{1+\omega^2\tau^2} \qquad \text{(C.7.8)}$$

At equilibrium the second term vanishes. The principle of detailed balance means

$$k_1' N_1^s = k_2'' N_2, \qquad k_o' N_o = k_1'' N_1^s \tag{C.7.9}$$

Therefore at equilibrium:

$$G_{\Delta J}(\omega) = 2\,A_1 - \frac{2\tau}{1+\omega^2\tau^2}\,N_1^s(\gamma_2 k_1' - \gamma_1 k_1'')^2 \tag{C.7.10}$$

At nonequilibrium the frequency-dependent contribution to $4\,k_bT_a\mathrm{Re}Y(\omega)$ may be negative or positive, depending on the response properties of the system. It is positive e.g. for the special case of pores under very asymmetric conditions

$$\gamma_1 > \gamma_2, \qquad k_1'' = k_2'' = 0, \tag{C.7.11}$$

where flux can take place only in one direction.

The second term in (C.7.8) can also be positive and negative, because the sign of $(\gamma_2 k_1' - \gamma_1 k_1'')$ may be equal with or different from the sign of J^s, depending on the special choice of the different parameters. Taking the Eyring ansatz (C.3.5) for the voltage-dependence of the rate constants the absolute value of this term becomes greater for higher applied voltages, if it is positive, and smaller, if it is negative. Generally, it is negative in cases where the gradient between concentrations at the left and right pore-side is opposite to the direction of the applied voltage.

The results for open channels with one binding site have illustrated some general properties of transport fluctuations. As mentioned above the property of the spectral density at equilibrium to be smaller at low frequencies than at high frequencies is a direct consequence of the fact that the linear current response (for $t > o$) is in opposite direction to the applied voltage pulse (at $t = o$). This indicates the impossibility for the occurrence of current excess noise as e.g. $1/f$ noise or nerve channel $(1/f^2)$ noise at (thermal) equilibrium. Nevertheless the process itself, which is the actual source of the nonequilibrium excess noise, may be an equilibrium process, as for example the opening and closing of nerve channels, generating $1/f^2$-current noise.

At nonequilibrium under special conditions the current response may be in the same direction as the applied voltage pulse and hence yield an excess noise contribution of the admittance term to the fluctuations.

The bilinear dependence on current of the second term in (C.7.4) may often result in a quadratic dependence, which is typical of most excess noise sources.

The fact that the sign of both contributions to the fluctuations can be negative for a suitable choice of parameters might include interesting aspects for the problem of reduction of noise in special frequency intervals in amplifiers as well as in biological nonequilibrium transport. This will be subject of discussion in section 8.

7.2 Different Behaviour of Scalar and Vectorial Quantities at Nonequilibrium

We believe that our concept of transport in discrete systems offers advantages compared with other approaches to transport. The discrete description seems to be adequate for systems with discontinous structures and a coupling with other processes as e.g. chemical reactions. An example for the latter case is the discussed simple model of carrier mediated ion transport. Even models of ion transport through pores with 'memory' effects may be treated where the pores may assume different conformational states with different conductivities (see section 6 or Läuger et al., 1980), which may depend on the ionic occupational states of the pores. These and other situations as e.g. the single-file transport through narrow pores can be described in a satisfactory way only with the discrete transport concept. Especially in systems, where the decisive transport processes take place in microscopic dimensions comparable with the dimensions of the transported stuff, a discrete concept is more appropriate than continuous models. Obviously, in principle there are no difficulties concerning the approximation of continuum models by discrete systems in an appropriate limit. As consequence the general results derived e.g. for the nonequilibrium transport fluctuations, can be applied also for continuum transport processes.

In this connection it will be important to investigate the general range of validity of the results. This does not only concern the problem of approximation of continuum systems but also the question if and how non-Markovian systems might be treated as Markovian systems by a suitable increase in the number of states of the systems. We believe that in this way the discrete concept may be used very generally and hence the following points of discussion have a more general range of validity.

Much work concerning the treatment of nonequilibrium steady states has been done. In a classical paper Lax (1960) has derived a generalization of the Nyquist or fluctuation-dissipation theorem for nonequilibrium steady states. Since that time and the important work of Van Vliet and Fasset (1965) further papers concerning the treatment of steady state fluctuations, which are based on the so-called master equation approach have been published. Keizer (1976) has introduced a state function matrix $\underset{\sim}{\sigma}$ (essentially the inverse of the variance matrix) as a generalization of entropy. Chen (1978) has used the master equation approach e.g. for the treatment of (nonequilibrium) electrical nerve channel fluctuations and similarly as Jähnig and Richter (1976) has proposed a generalized Nyquist relation for steady states. Apart from a factor containing k_b and T_a the complex admittance matrix is introduced by

$$\underset{\sim}{Y}(\omega) \propto (\underset{\sim}{M} - \frac{i}{\omega} \underset{\sim}{\sigma}^2)^{-1} \qquad (C.7.10)$$

which immediately by comparison with the noise spectrum matrix (B.3.28) leads to an extension of the fluctuation-dissipation theorem to nonequilibrium steady states for the spectral density matrix $\underset{\sim}{G}_{\dot{\underset{\sim}{\alpha}}}(\omega)$ of the 'fluxes' $\dot{\underset{\sim}{\alpha}}=\frac{d\underset{\sim}{\alpha}}{dt}$.

There seems to be a discrepancy with our result showing the breakdown of the fluctuation-dissipation theorem at non-equilibrium. But this discrepancy can be clarified in the following way: The mentioned treatments of steady states all are concerned with scalar fluxes, which are given by the time derivatives of the concentrations (state variables), while our treatment of directed transport is based upon fluxes of vectorial character, formed by the transitions between different states. With the definition (C.2.4) of divergence in discrete systems, the time derivatives of the state variables are given according to the balance (C.2.5) by the divergence of fluxes in analogy to the continuity equations for the vectorial fluxes of mass, charge or heat in continuum diffusion processes:

$$\text{div}\ \underset{\sim}{\phi}(\nu)=\dot{N}_{\nu} \qquad (C.7.11)$$

It is well known (e.g. Casimir, 1945, De Groot a. Mazur, 1962, Meixner a. Reik, 1959) that at equilibrium vectorial (and tensorial) fluxes (and forces) behave as scalar quantities, e.g. satisfy the Onsager-Casimir reciprocity relations or the fluctuation-dissipation theorem. In agreement with these findings for an appropriate choice of the external force our results show the validity of the fluctuation-dissipation theorem at equilibrium states. Thus a main result is the essentially different behavior of scalar and vectorial quantities at nonequilibrium. While at equilibrium fluctuation-dissipation relations are valid for both types, at non-equilibrium such relations no longer hold for transport observables based on vectorial fluxes. The transport fluctuations are built up by two contributions, one being related to the admittance $Y_p^T(\omega)$ and the other with bilinear structure in the vectorial (directed) fluxes. This second term yields an essentially different type of fluctuations, which is coupled to the directed character of vectorial quantities and does not occur in the fluctuations of scalar quantities. For electrical fluctuations this type is responsable for the so-called excess noise phenomena, e.g. 1/f-noise or $1/f^2$-channel noise.

The problem, under which special conditions the current (or another transport observable T) can be represented by a linear combination of the time derivatives of the state variables and hence a fluctuation-dissipation relation with (C.7.10) may be found, can be uniquely answered with the theorem proven in section 2.3: This is possible only for systems which are conservative with respect to T (rot $\tilde{\gamma}$=o), and then according to (C.2.15) the stationary value T^s vanishes. Hence it is possible only under the necessary ('equilibrium') condition T^s=o and always impossible for nonvanishing steady state values of the transport observable.

8. Current Noise: The Limit of High Applied Voltage

In section 2.3 of part B and in section 1.2 of part C we have discussed simple examples of transport noise, which in the limit of high applied voltage and for low frequencies (low compared with the transition rates) resulted in the usual shot-noise relation for the spectral density

$$G_{\Delta J}(\omega) = 2\, z\, e_0\, J^s \qquad \text{(C.8.1)}$$

In the extreme nonequilibrium-situation of high applied voltage the transitions(jumps) between two states (sites) are assumed to take place only in one direction. We will show that under this condition in the case of negligible interactions the shot-noise relation may be derived very generally with the use of the presented approach to transport fluctuations in discrete systems. At the special example of interactions in the single-file transport model it is demonstrated that the intensity of the fluctuations is strongly influenced by the interactions. Hence the validity of the shot-noise relation is restricted to the case of independent events.

8.1 Interactions Neglected: Shot Noise

If interactions in the transport system are neglected we can start from the linear phemomenological equations(C.2.16). We first consider the strictly one-dimensional case where only one ionic pathway can be used by the ions for crossing the transport system between two reservoirs. The total voltage is assumed to be applied between these reservoirs. Hence, if only one kind of charged units (charge $z\,e_o$) is involved in the process, the sum over the corresponding $\gamma_{\alpha\beta}$ equals $z\,e_o$. Under these assumptions for all $\phi^s_{\mu\nu}$ holds

$$\phi^s_{\mu\nu} = \phi^s, \ \phi^s_{\nu\mu} = o \qquad (C.8.2)$$

or reverse. ϕ^s is the constant stationary flux through the one-pathway transport system. We number the sites ν along the pathway in the direction of the applied voltage. Then from the definition of the fundamental solutions (B.3.16a) ($\tilde{Y}$≠o!) follows

$$M_{\kappa\rho} \int_0^\infty \Omega_{\rho\mu}(t)\,dt = \begin{cases} 1 & \text{for } \mu \le \rho, \rho = \kappa+1 \\ o & \text{otherwise} \end{cases} \qquad (C.8.3)$$

If double jumps $\nu \rightarrow \nu+2$ or higher order jumps are admitted, sequencies with these jumps are treated as different ionic pathways and hence have to be excluded in the one-pathway-case.

With $ze_o\phi^s = J^s$ we get from (C.4.5) for $\omega \to o$

$$G_{\Delta J}(o) = 2\phi^s \sum_{\substack{\mu,\nu \\ \nu=\mu+1}} \gamma^2_{\mu\nu} + 4\phi^s \sum_{\substack{\mu,\nu,\kappa,\rho, \\ \nu=\mu+1 \\ \rho=\kappa+1 \\ \mu\le\rho}} \gamma_{\mu\nu}\gamma_{\kappa\rho} =$$

$$= 2\phi^s \Big(\sum_{\substack{\mu,\nu \\ \nu<\mu}} \gamma_{\mu\nu}\Big)^2 = 2z^2 e_o^{\,2} \phi^s$$

Hence the shot-noise relation

$$G_{\Delta J}(o) = 2\,z\,e_o\,J^s \qquad (C.8.4)$$

is derived in the low frequency limit.

This result may easily be extended to cases where the transported particles can cross the system through different pathways. The i-th pathway is treated as above with the fraction $\phi^s_{(i)}$ of stationary flux going through this pathway and

$$\sum_i \phi^s_{(i)} = \phi^s \qquad (C.8.5)$$

Summation over the contributions to $G_{\Delta J}(o)$ from all pathways yields (C.8.4), which therefore is proven to hold generally in the case of negligible interactions and one species of charged units being involved in the transport process which generates the electric current. Generalization to cases where different kinds of charged units generate the current can be done without problem under the condition of negligible interaction.

Finally, we can ask in which way the admittance term and the excess-noise term contribute to shot noise. With (C.8.2), (C.3.17) and (C.4.6) one gets e.g. in the one-pathway case for the admittance

$$4\, k_b T_a \mathrm{Re}\, Y(\omega) = z\, e_o\, J^s + \phi^s \sum_{\substack{\mu,\nu \\ \nu<\mu}} \gamma^2_{\mu\nu} \qquad (C.8.6)$$

and for the excess -noise term

$$G_{\Delta J}(\omega) - 4\, k_b T_a \mathrm{Re}\, Y(\omega) = z\, e_o\, J^s - \phi^s \sum_{\substack{\mu,\nu \\ \nu<\mu}} \gamma^2_{\mu\nu} \qquad (C.8.7)$$

Thus shot noise is built up by both terms.

8.2 Reduction of Noise as Consequence of Ionic Interactions in the Single-File Model

In case interactions essentially determine the transport process the shot noise relation can no longer be expected to be valid. Indeed, by first simple model calculations of noise generated by the single-file transport of ions through narrow pores (one-ion case) we could show that the (low-frequency) noise intensity is reduced below the shot noise level with the magnitude of reduction depending on the number of binding sites in the channels (Frehland, 1981).
We believe that the dependence of the low frequency noise intensity on special interaction mechanisms should be subject of detailed future investigation for the following reasons:

1. One can speculate if and how biological nonequilibrium transport systems during their evolution have reached states of minimum fluctuations. 2. The possibilities of application to the construction of amplifiers of low noise level should beinvestigated. 3. The experimental measurement of low-frequency current noise intensities could be a possibility of getting information about the transport parameters from fluctuation analysis though the transport times are too fast and the corresponding dispersive frequency regions too high to be measurable.

The following analytical and numerical results for single-file noise will demonstrate that the presented approach to transport fluctuations in discrete systems is suitable for investigation of these problems. First we determine the noise intensity for high applied voltage of single-file noise in pores with regular barrier structure in the case where the pores are to be occupied by not more than one ion (one-ion case). The corresponding state diagram for a single pore with n binding sites is shown in Fig. C17. All rate constants and the corresponding $\gamma_{\mu\nu}$ are assumed to be equal:

$$\gamma_{\mu\nu} = \frac{ze_0}{n+1} \qquad (C.8.6)$$

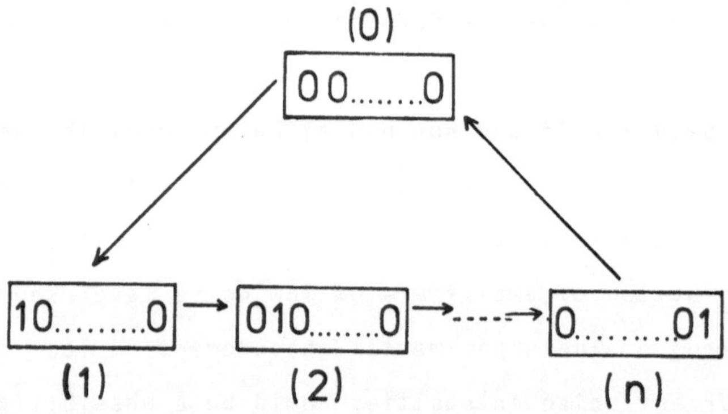

Fig. C17: One-ion state diagram for pores with regular barrier structure and n binding sites in the case of high applied voltage.

We start from the master equation for the single channel. It is easy to show that the frequency-dependent second term in (C.4.5) vanishes and hence the current noise spectrum is white. Taking into account that

$$M_{\kappa\rho}\ \Omega_{\rho\mu}(t)$$

is the deviation of flux $\phi_{\kappa\rho}$ from the steady state (under the special condition distinguishing state μ), the sum over all fluxes

$$\sum_{\kappa,\rho} M_{\kappa\rho}\Omega_{\rho\mu}(t)$$

vanishes, because the structure is regular, and negative and positive terms cancel. Therefore with the equality of all $\gamma_{\kappa\rho}$ also

$$\int_0^\infty \sum_{\kappa,\rho} \gamma_{\kappa\rho} M_{\kappa\rho}\Omega_{\rho\mu}(t)\cos\omega t = 0 \qquad (C.8.9)$$

and the white current noise spectrum is with (C.4.5) and (C.2.9)

$$G_{\Delta J}(\omega) = \frac{1}{n+1}\ 2\ z\ e_0\ J^s \qquad (C.8.10)$$

The result clearly shows that the spectral density is below the shot noise level by the factor $\frac{1}{n+1}$. For one binding site it is half the shot noise intensity. With increasing number of sites the noise intensity is further reduced.

In the figures C18 and C19 we present a number of numerical calculations of current noise in the complete single-file model for pores with two binding sites (C18) and three binding sites (C19). As to be seen from the legends to both figures

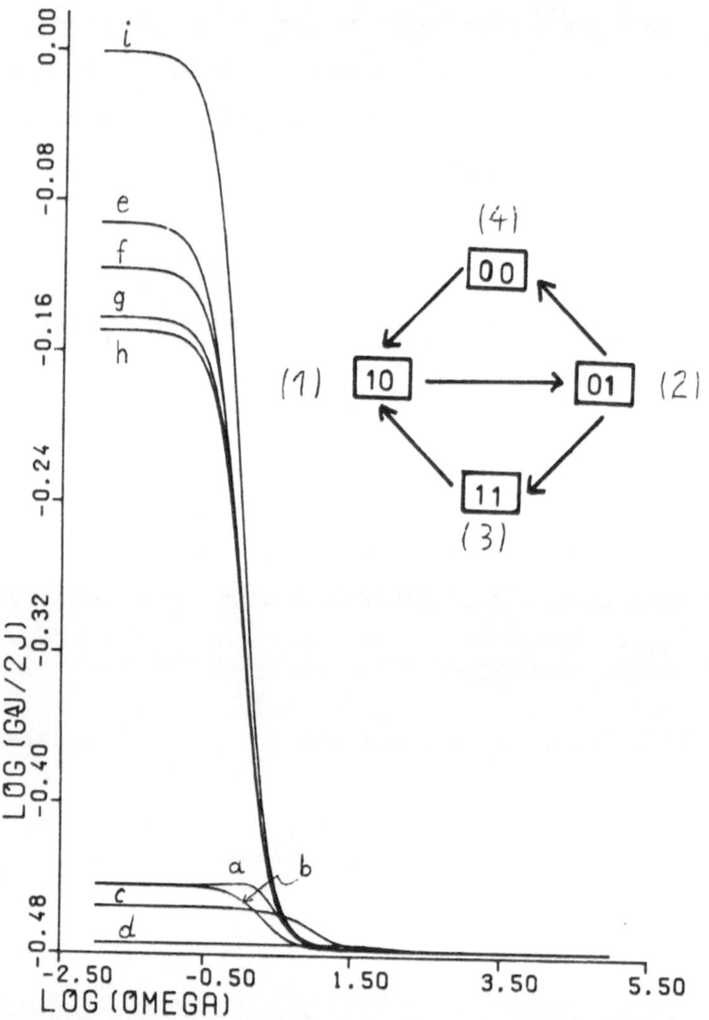

Fig.C18: Single-file current noise for high applied voltage in channels with two binding sites. All $\gamma_{\mu\nu}$ are set equal to ½. The rate rate constants $M_{\mu\nu}$ ($\nu \neq \mu$) for the different curves are zero with the exceptions

a) $M_{14}=1$, $M_{21}=1$, $M_{42}=1$, $M_{13}=1$, $M_{32}=1$

b) as a) except $M_{32}=\frac{1}{2}$, $M_{13}=2$

c) as a) except $M_{32}=\frac{1}{10}$, $M_{13}=10$

d) as a) except $M_{32}=\frac{1}{100}$, $M_{13}=100$ (one-ion case)

e) $M_{14}=\frac{1}{10}$, $M_{32}=\frac{1}{10}$, $M_{13}=M_{21}=M_{42}=1$

f) as e) except $M_{13}=2$, $M_{32}=\frac{1}{20}$

g) as e) except $M_{13}=10$, $M_{32}=\frac{1}{100}$

h) as e) except $M_{13}=100$, $M_{32}=\frac{1}{1000}$

i) as h) except $M_{14}=\frac{1}{1000}$ (no-interaction case)

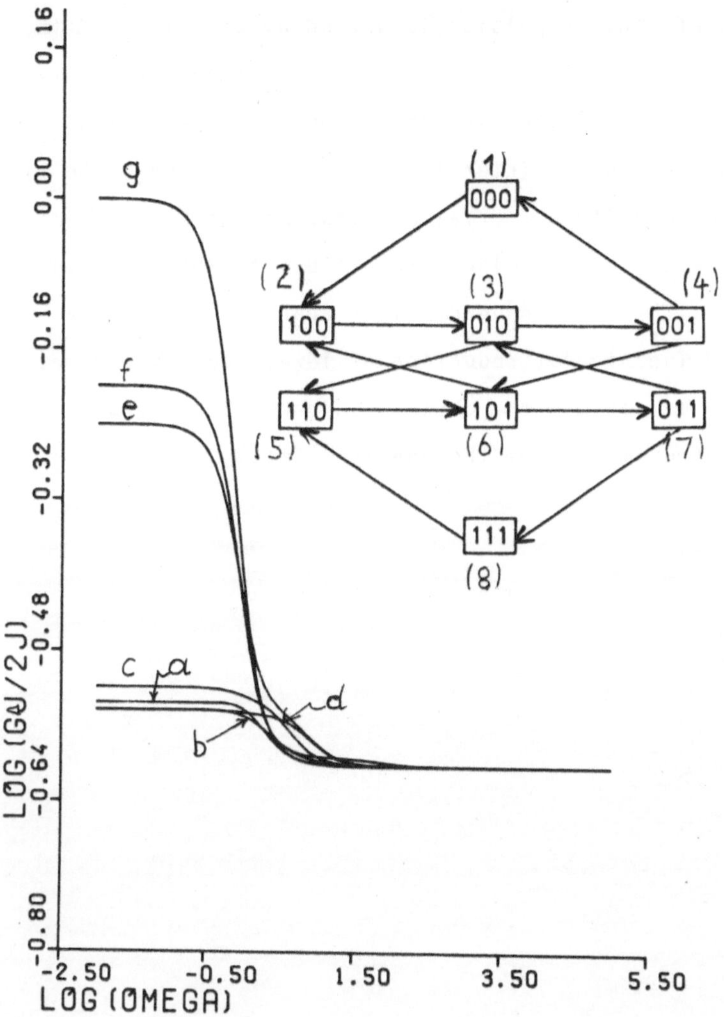

Fig.C.19: Single-file current noise for high applied voltage in channels with three binding sites. All $\gamma_{\mu\nu}$ are set equal to $\frac{1}{4}$. The rate constants $M_{\mu\nu}$ $(\nu \neq \mu)$ for the different curves are zero with the exceptions

a) $M_{21}=M_{14}=M_{32}=M_{43}=M_{26}=M_{37}=M_{53}=M_{64}=M_{65}=M_{76}=M_{87}=M_{58}=1$

b) as a) except $M_{87}=\frac{1}{10}$, $M_{58}=10$ (weak 3-ion cycle)

c) as a) except $M_{87}=\frac{1}{100}$, $M_{58}=100$ (≈ no 3-ion cycle)

$M_{37}=M_{26}=M_{65}=5$, $M_{76}=M_{64}=\frac{1}{5}$(weak 2-ion cycles)

e) as d) except $M_{21}=\frac{1}{5}$ (weak 1-ion cycle)

f) $M_{87}=M_{76}=M_{64}=M_{53}=\frac{1}{100}$; $M_{58}=M_{65}=M_{26}=M_{37}=100$

$M_{32}=M_{43}=M_{14}=1$, $M_{21}=\frac{1}{100}$ (weak interactions)

g) as e) except $M_{21}=M_{64}=M_{53}=\frac{1}{1000}$ (no-interaction case)

the degree of ionic repulsion is varied by variation of the corresponding transition rate constants. The non-interaction case is also approximated in both examples. By division through the shot noise intensity $2J^S$(ze_o is set equal to 1) the spectral density is normalized thus making possible a sensible comparison of noise intensities. In the non-interaction cases the low frequency limit approximates 1, i.e. shot noise. The highest reduction of low-frequency current noise is to be seen in the cases of highest ionic repulsion, i.e. the approximation of the one-ion cases.

D. Nonstationary Processes, Fluctuations at Transient States

In the last years the electric fluctuations generated by ionic channels in nerve membranes at nonstationary states have been measured and analyzed (Sigworth, 1977, 1981).
In this part we discuss some aspects of theoretical analysis, mainly based on time dependent equations for the moments, which can be derived from the master equation (B.3.6). There exists an extensive literature concerning these problems in chemical kinetics, population dynamics or other fields (see e.g. van Kampen, 1975, Goel a. Richter - Dyn, 1974, Mc Quarrie, 1967, Dublin, 1976). Furthermore we shortly describe how the approach to transport fluctuations around steady states in part C can be extended to transient states.

1. A Simple Example: Identical, Independent Channels

If stochastic fluctuations are generated by a number of completely independent and identical subunits, which may be in two alternative states, the stochastic analysis of transient states may be done by simple application of the binomial law. For example, if a membrane transport system consists of N independent identical channels, which may be alternatively open (with a single channel conductance Λ) or closed, then the probability $P_N(N_{op})$ to find N_{op} channels in the open state is given by the binomial law (A.3.1). For N_{op} may be considered as the outcome of N independent identical experiments. Then for a probability P of the single channel to be open $P_N(N_{op})$ is according to (A.3.1):

$$P_N(N_{op}) = \binom{N}{N_{op}} P^{N_{op}}(1-P)^{N-N_{op}} \qquad (D.1.1)$$

If the transport system is in a transient state (e.g. after a voltage jump) the probability P and hence also $P_N(N_{op})$ is time dependent

$$P = P(t) \qquad (D.1.2)$$

but (D.1.1) still holds. Hence (D.1.1) may be used for a description of fluctuations at transient states. The first moment, i.e. the expectation (mean) value $\langle N_{op} \rangle$ and the variance σ^2 are according to (A.3.2)

$$\langle N_{op}(t) \rangle = N \cdot P(t) \qquad (D.1.3)$$

$$\langle \sigma^2(t) \rangle = N \cdot P(t)(1-P(t)) = \langle N_{op}(t) \rangle \; (1-P(t))$$

(D.1.3) may be used for the analysis of current fluctuations generated by the opening and closing of the channels under the assumption that the transport kinetics are fast compared with open-closed kinetics (c.f. the discussion of this point in chapters (B.3.7) and (C.6.3)). Then the measured current J(t) is determined by the applied voltage V, single channel conductance Λ and number of open channels and wie get from (D.1.3) the relations

$$J = V \cdot \Lambda \cdot N \cdot P$$

$$\sigma_J^2 = V^2 \Lambda^2 NP(1-P) \qquad (D.1.4)$$

$$= V \Lambda J - \frac{1}{N} J^2$$

Thus a measurement of J(t) and $\sigma_J^2(t)$ can be used for testing the applicability of the binomial law or for determination of the single channel conductance Λ and the total number of acting channels.

2. Time-Dependent States in Markov-Processes

In this and the following chapters we describe an approach to the analysis of fluctuations in time-dependent states based on the master equation in the form (B.3.6). Mainly we shall discuss the time-dependent equations of the first two moments.

2.1 General Equations for the Moments

In section (B.3.6) we have shortly described how to arrive at general equations for the moments. Starting with the master equation (B.3.6)

$$\frac{d P(\underset{\sim}{N}, t/\underset{\sim}{N}'; o)}{d t} = \sum_{\underset{\sim}{N}''} P(\underset{\sim}{N}'', t/\underset{\sim}{N}'; o) \cdot Q(\underset{\sim}{N}; \underset{\sim}{N}'')$$

$$- \sum_{\underset{\sim}{N}''} P(\underset{\sim}{N}, t/\underset{\sim}{N}'; o) \cdot Q(\underset{\sim}{N}''; \underset{\sim}{N}) \quad , \qquad \text{(D.2.1)}$$

the first moment or expectation value of $\underset{\sim}{N}$ is defined by (c.f.(A.2.10) and (B.3.8)):

$$<\underset{\sim}{N}(t)>_{\underset{\sim}{N}(o)} = \sum_{\text{all } \underset{\sim}{N}} \underset{\sim}{N} \cdot P(\underset{\sim}{N}, t/N(o)) \qquad \text{(D.2.2)}$$

Correspondingly the higher moments:

$$\underset{\sim}{N}^k(t)_{\underset{\sim}{N}(o)} = \sum_{\text{all } \underset{\sim}{N}} \underset{\sim}{N}^k \cdot P(\underset{\sim}{N}, t/N(o)) \qquad \text{(D.2.3)}$$

With (D.2.2) and (D.2.3) general equations for the time-dependence of the moments may be derived from (D.2.1).

E.g. for the first moment the result is (c.f.(B.3.9)):

$$\frac{d \langle \underset{\sim}{N}(t) \rangle_{\underset{\sim}{N}(o)}}{d\, t} = \sum_{\text{all } \underset{\sim}{N}''} \underset{\sim}{A}(\underset{\sim}{N}'') P(\underset{\sim}{N}''; t/N(o)) = \langle \underset{\sim}{A}(\underset{\sim}{N}) \rangle_{\underset{\sim}{N}(o)} \tag{D.2.4}$$

with the first order Fokker-Planck moments

$$\underset{\sim}{A}(\underset{\sim}{N}'') = \sum_{\text{all } \underset{\sim}{N}} (\underset{\sim}{N}-\underset{\sim}{N}'') Q(\underset{\sim}{N};\underset{\sim}{N}'') \quad . \tag{D.2.5}$$

Correspondingly equations for the higher moments are derived.

The treatment of the general case, where the state vector $\underset{\sim}{N}$ represents a set of variables, is rather complex. In the following we shall restrict to the so-called one-variable birth-death processes. The main problems concerning the solution of the equations for the moments may be discussed at these systems.

2.2 One-Variable-Birth-Death Processes

We define the one-variable birth-death (generation-recombination) process as a Markov process which is discrete in space and homogeneous in time with the following special properties:

a) The random variable can assume only integer values
$\ldots, -2, -1, 0, +1, +2, \ldots$.

b) Transitions are possible only between neighbouring states:
$n \to n+1$ (birth or generation),
$n \to n-1$ (death or recombination)

It is useful to introduce the following simplified notation for the conditional probabilities and transition moments

$$P_{n,m}(t) = P(x=n, t/x'=m, o) \tag{D.2.6}$$

$$\lambda_n = Q(n+1;n) \tag{D.2.7}$$

$$\mu_n = Q(n-1;n) \tag{D.2.8}$$

Then the master equation (D.2.1) for the one-variable birth-death process may be simplified to

$$\frac{d\,P_{n,m}(t)}{dt} = \lambda_{n-1}\,P_{n-1,m}(t) - (\mu_n + \lambda_n) P_{n,m}(t) + \mu_{n+1}\,P_{n+1,m}(t) \tag{D.2.9a}$$

The solutionsof (D.2 9a) satisfy the initial conditions

$$P_{n,m}(o) = \delta_{nm} \qquad (D.2.10)$$

(δ_{nm}: Kronecker delta). Having solved (D.2.9a) for arbitrary m the general solution for the probability $P_n(t)$ with an arbitrary initial probability distribution

$$P_n(o) = \overset{o}{P}_m \quad , \quad \sum_m \overset{o}{P}_m = 1 \qquad (D.2.11)$$

is simply given by these fundamental solutions through

$$P_n(t) = \sum_m P_{n,m}(t) \overset{o}{P}_m \qquad (D.2.12)$$

Because of the linearity of the master equation (D.2.9a) $P_n(t)$ with arbitrary initial condition satisfies an equation of the form (D.2.9a)

$$\frac{d P_n(t)}{dt} = \lambda_{n-1} P_{n-1}(t) - (\mu_n + \lambda_n) P_n(t) + \mu_{n+1} P_{n+1}(t) \qquad (D.2.9b)$$

In (D.2.9a) the process is initially in a definite state (e.g. $\sigma^2(o) = o$ in this case!), while in (D.2.9b) an initial probability distribution is allowed. We stress this point , because in many experimental situations the initial state is not well defined. See e.g. below in section (D.4.3) the application to a bimolecular reaction, where the variance $\sigma^2(o)$ at t=o must be assumed to benonzero

Some remarks should be made concerning the boundary conditions, which depend on the allowed set of states of the random variable. In principle there may be distincted

i) processes without restrictions (in n), i.e. n may assume all integer values, and

ii) processes with restrictions for the range of possible values in n.

In the second case two important boundary conditions are given by so-called reflecting or absorbing barriers. In the case of a reflecting barrier at N the variable cannot assume values greater than N. If N is an absorbing barrier, the system remains in state N, if it once has reached this state. For example the number of species of a population in an isolated region has an absorbing barrier at N=o (distinction). For further details the reader is referred to the literature (e.g. Goel, Richter-Dyn, 1974).

In the examples, which we shall discuss below, reflecting barriers are automatically taken into account by the special structure of the transition moments.
As described above we may derive from the master equation (D.2.9) differential equations for the moments. Multiplying (D.2.9) with n and summation over n yields the following equation for the expected value $\langle n \rangle$

$$\frac{d\langle n\rangle(t)}{dt} = \langle \lambda_n \rangle(t) - \langle \mu_n \rangle(t) \qquad \text{(D.2.13)}$$

with

$$\langle \lambda_n \rangle(t) = \sum_{\text{all } n} \lambda_n P_n(t) \quad , \quad \langle \mu_n \rangle(t) = \sum_{\text{all } n} \mu_n P_n(t) \qquad \text{(D.2.14)}$$

For derivation of (D.2.13) we have used that

$$\sum_{\text{all } n} n\,\lambda_{n-1}\,P_{n-1}(t) = \sum_{\text{all } n} (n+1)\,\lambda_n\,P_n(t)$$

and correspondingly

$$\sum_{\text{all } n} n\mu_{n+1}\,P_{n+1}(t) = \sum_{\text{all } n} (n-1)\,\mu_n\,P_n$$

In a similar way for the k-th moments $<n^k>$ hold the equations

$$\frac{d <n^k>(t)}{dt} = < \{ (n+1)^k - n^k \} \lambda_n > - < \{ n^k - (n-1)^k \} \mu_n > \quad \text{(D.2.15)}$$

Then easily follows an equation for the time dependent variance $\sigma^2(t)$:

$$\frac{d\,\sigma^2(t)}{dt} = 2\,\{<n\lambda_n> - <n\mu_n>\} + (1-2<n>)<\lambda_n> + (1+2<n>)<\mu_n> \quad \text{(D.2.16)}$$

For derivation of (D.2.13) - (D.2.16) no special initial conditions (D.2.10) have been used. They are valid also for initially distributed states $\overset{o}{P}_m$. As we shall see the difficulties concerning the solutions of these equations are enormeous with the exception of linear processes. For in general the equation (D.2.13) for the time dependence of the first moment involves (in $<\lambda_n>$ and $<\mu_n>$) the second moment and so on. Below we shall discuss the problem how to solve this hierarchy of equations approximately.

3. Linear Birth-Death Processes

In case the transition probabilities λ_n and μ_n are linear in n, we call the process 'linear', otherwise it is nonlinear. In the linear case the equation for the k-th moment involves the moment only up to the k-th moment. Hence the hierarchy of equations for the moments can be solved step by step. In the following we shall discuss explicitely the equations for the mean value $\langle n \rangle$ and the variance σ^2.

3.1 Expectation Value and Variance

In a linear process the transition moments are linear in n:

$$\lambda_n = \overset{0}{\lambda} + \overset{1}{\lambda} \cdot n, \quad \mu_n = \overset{0}{\mu} + \overset{1}{\mu} \cdot n \qquad \text{(D.3.1)}$$

Then the expectation values $\langle \lambda_n \rangle$ and $\langle \mu_n \rangle$ are

$$\langle \lambda_n \rangle = \overset{0}{\lambda} + \overset{1}{\lambda} \langle n \rangle \qquad \text{(D.3.2)}$$

$$\langle \mu_n \rangle = \overset{0}{\mu} + \overset{1}{\mu} \langle n \rangle$$

λ_n and μ_n must be non-negative for all allowed values of n. This does not mean that all parameters $\overset{0}{\lambda}$, $\overset{1}{\lambda}$, $\overset{0}{\mu}$ and $\overset{1}{\mu}$ are non-negative.

With (D.3.2) the differential equation for the first moment reads

$$\frac{d \langle n \rangle}{dt} = (\overset{0}{\lambda} - \overset{0}{\mu}) + (\overset{1}{\lambda} - \overset{1}{\mu}) \langle n \rangle \qquad \text{(D.3.3)}$$

and for the variance

$$\frac{d\sigma^2}{dt} = 2\,(\overset{1}{\lambda}-\overset{1}{\mu})\sigma^2 + (\overset{1}{\lambda}+\overset{1}{\mu})\langle n\rangle + (\overset{o}{\lambda}+\overset{o}{\mu}) \qquad (D.3.4)$$

On the contrary to nonlinear processes the equation for the first moment does not depend on the variance. The fluctuations do not influence its time course. Hence in linear processes the macroscopic (deterministic) equations describe also the time dependence of the first moment.

The linear equations (D.3.3) and (D.3.4) are solved by standard methods. One gets

$$\langle n\rangle(t) = \langle n\rangle_o\, r + \frac{(\overset{o}{\lambda}-\overset{o}{\mu})\,\gamma}{\overset{1}{\lambda}\,(\gamma-1)}\cdot(r-1) \qquad (D.3.5)$$

with

$$\langle n\rangle_o = \langle n\rangle(o),$$

$$r = e^{(\overset{1}{\lambda}-\overset{1}{\mu})t}, \qquad \gamma = \overset{1}{\lambda}/\overset{1}{\mu} \qquad (D.3.6)$$

and

$$\sigma^2(t) = \frac{\langle n\rangle_o\, r(r-1)(\gamma+1)}{\gamma-1} + \frac{(\overset{o}{\lambda}-\overset{o}{\mu})\,\gamma(r-1)(\gamma r-1)}{\overset{1}{\lambda}(\gamma-1)^2} +$$

$$+\ \sigma_o^2\, r^2 \qquad (D.3.7)$$

with $\sigma_o^2 = \sigma^2(o)$. The time dependence of $\langle n\rangle$ is described by one relaxation time $\tau = (\overset{1}{\lambda}-\overset{1}{\mu})^{-1}$, while $\sigma^2(t)$ is described by the two relaxation times τ and $\frac{1}{2}\tau$.

3.2 Application: Linear Two State Channel Models

As a simple test the above results may be applied for a description of channel fluctuations at transient states, if the channels act independently and have two states (open and closed):

$$\text{open} \underset{k_{op}}{\overset{k_{cl}}{\rightleftarrows}} \text{closed} \qquad \text{(D.3.8)}$$

In this case the results must be in agreement with the treatment in section 1 based on the binomial law. The open-closed kinetics are described by the equations

$$\frac{d\langle N_{op}\rangle}{dt} = -k_{cl}\langle N_{op}\rangle + k_{op}\langle N_{cl}\rangle = -\frac{d\langle N_{cl}\rangle}{dt} \qquad \text{(D.3.9)}$$

N_{op}: number of channels in the open state;
N_{cl}: number of channels in the closed state .

If N is the (constant) total number of channels,

$$N_{op} + N_{cl} = N \qquad \text{(D.3.10)}$$

we get from (D.3.9)

$$\frac{d\langle N_{op}\rangle}{dt} = -k_{cl}\langle N_{op}\rangle - k_{op}\langle N_{op}\rangle + k_{op}N \qquad \text{(D.3.11)}$$

In order to treat the fluctuations, we now must associate the parameters in (D.3.11) to the parameters $\overset{o}{\lambda}$, $\overset{1}{\lambda}$, $\overset{o}{\mu}$, $\overset{1}{\mu}$, which characterize the birth-death transition. A solution of equation (D.3.11) alone which could describe also processes with a completely different stochastic behaviour, does not give information about the fluctuations. Obviously the first term on the right-hand side of (D.3.9) describes the 'death' transitions, and the second term the 'birth' transitions, where the opening (closing) of a channel is considered as a birth (death) transition. Hence the parameters $\overset{i}{\lambda}$, $\overset{i}{\mu}$ are with (D.3.11)

$$\overset{0}{\mu} = 0 \quad , \quad \overset{1}{\mu} = k_{cl} \quad , \quad \overset{0}{\lambda} = k_{op}\cdot N \quad , \quad \overset{1}{\lambda} = -k_{op} \tag{D.3.12}$$

For the allowed values of N_{op} $(0 \leq N_{op} \leq N)$ λ_n and μ_n are positive though $\overset{1}{\lambda}$ is negative.

Now regard the transient state from a stationary state at t = 0 with rate constants $k_{o,op}$, $k_{o,cl}$ to a stationary state (for $t \rightarrow \infty$) with k_{op}, k_{cl}. The total number of channels N is considered to be constant. The application of (D.3.5) (D.3.7) yields the following results for the initial state:

$$\langle N_{op} \rangle_o = N\cdot \frac{k_{o,op}}{k_{o,op}+k_{o,cl}} \tag{D.3.13}$$

$$\sigma_o^2 = N\cdot \frac{k_{o,op}\cdot k_{o,cl}}{(k_{o,op}+k_{o,cl})^2} \tag{D.3.14}$$

And for the transient state:

$$\langle N_{op} \rangle (t) = \langle N_{op} \rangle_o \, r(t) + \langle N_{op} \rangle_\infty \, (1-r(t)) \tag{D.3.15}$$

$$\sigma^2(t) = \langle N_{op} \rangle_o \, r(t) \, (1-r(t))\cdot\left(\frac{k_{cl}-k_{op}}{k_{cl}+k_{op}}\right) + \sigma_\infty^2 \, (1-r(t)) \left(1+ \frac{k_{op}}{k_{cl}} r(t)\right) + \sigma_o^2\cdot r^2(t) \tag{D.3.16}$$

with

$$r(t) = e^{-(k_{op}+k_{cl})\cdot t} \qquad (D.3.17)$$

and the final state

$$<N_{op}>_{\infty} = N \frac{k_{op}}{k_{op}+k_{cl}} \qquad (D.3.18)$$

$$\sigma_{\infty}^{2} = N \frac{k_{op}\ k_{cl}}{(k_{op}+k_{cl})^{2}} \quad . \qquad (D.3.19)$$

These relations should agree with the results in chapter 1 which have been obtained by application of the binomial distribution. Indeed, with the probability

$$P(t) = \frac{<N_{op}>(t)}{N} \qquad (D.3.20)$$

for the single channel to be open at time t, we get from relation (D.3.16) after some lengthy manipulations

$$\sigma^{2}(t) = N\ P(t)\ (1-P(t)) \qquad (D.3.21)$$

in agreement with (D.1.2).

We finally emphasize that the above results may be applied to a number of different chemical, physical or biological processes. Examples are: the unimolecular chemical reaction A $\leftrightarrow$ B, two level semiconductor models and transport of particles between two sites in a closed system (closed pores, simple gating models).

4. Nonlinear Processes

4.1 Approximation Procedures

As emphasized the equations (D.2.13) - (D.2.16) for the time dependence of the moments are solvable step by step only in the linear case. In the special example of linear birth-death processes the explicit equations for the first moment and the variance are given by (D.3.3) and (D.3.4). In nonlinear processes already the equation for the first moment involves higher moments(see below) and a set of coupled equations for the moments must be solved approximately. Only in special cases exact solutions of the master equation are available.

We do not want to go into a discussion of approximation procedures which have been proposed (see e.g. van Kampen 1975, Mc Quarrie, 1968, Goel a. Richter-Dyn, 1974, and references cited there). One class of procedures is based on the experience that for great numbers n the macroscopic equations mostly give a good description of the process, i.e. the stochastic deviations from the expectation value remain small. E.g. for a bilinear process (see section 4.2) the variance σ^2 occuring in the equation for the first moment $\langle n \rangle$ is assumed to be negligible compared with $\langle n \rangle^2$ (c.f. σ^2 for the binomial and Poisson distribution according to (A.3.3) and (A.3.7)).

We now derive approximate equations for the moments which can be solved because the equation for the n-th moment contains moments up to the order n only. We make the following two assumptions:

1) The deviations from the expectation value $\langle n\rangle(t)$ caused by stochastic fluctuations remain small, so that a linearization of λ_n and μ_n around $\langle n\rangle(t)$ is sensible.

2) There exist differentiable functions $\lambda_{\langle n\rangle}$ and $\mu_{\langle n\rangle}$ making possible the following linearization around $\langle n\rangle$

$$\lambda_n \approx \lambda_{\langle n\rangle} + \lambda'_{\langle n\rangle} \cdot (n-\langle n\rangle)$$

$$\mu_n \approx \mu_{\langle n\rangle} + \mu'_{\langle n\rangle} \cdot (n-\langle n\rangle) \qquad \text{(D.4.1)}$$

with

$$\lambda_{\langle n\rangle} := \lambda(\langle n\rangle), \quad \mu_{\langle n\rangle} := \mu(\langle n\rangle),$$

$$\lambda'_{\langle n\rangle} := \frac{d\lambda_{\langle n\rangle}}{d\langle n\rangle}, \quad \mu'_{\langle n\rangle} := \frac{d\mu_{\langle n\rangle}}{d\langle n\rangle}$$

Naturally both assumptions can be valid only for sufficiently great numbers n. With (D.4.1) follows from (D.2.13) the approximation

$$\frac{d\langle n\rangle(t)}{dt} = \lambda_{\langle n\rangle(t)} + \lambda'_{\langle n\rangle(t)} \quad \langle(n-\langle n\rangle(t))\rangle$$

$$-\mu_{\langle n\rangle(t)} - \mu'_{\langle n\rangle(t)} \quad \langle(n-\langle n\rangle(t)\rangle$$

Because $\langle(n-\langle n\rangle(t))\rangle = 0$, we get

$$\frac{d\langle n\rangle(t)}{dt} = \lambda_{\langle n\rangle(t)} - \mu_{\langle n\rangle(t)} \qquad \text{(D.4.2)}$$

And for $\sigma^2(t)$ follows similarly

$$\frac{d\sigma^2(t)}{dt} = \lambda_{<n>(t)}+\mu_{<n>(t)}+2(\lambda'_{<n>(t)}-\mu'_{<n>(t)})\sigma^2 \qquad (D.4.3)$$

The equation (D.4.2) for the first moment contains only the functional dependence of λ and μ on the first moment. After having solved this equation, we can solve equation (D.4.3) for the time-dependent variance, where on the right-hand side now only the variance itself is unknown. Thus at least the first two moments can approximately be determined by (D.4.2) and (D.4.3). By some lengthy manipulation of (D.2.15) with the use of (D.4.1) we generally get for the k-th moment

$$\frac{d\langle n^k\rangle}{dt} = \sum_{\nu=1}^{k} \binom{k}{\nu} \Big[\langle n^{k-\nu}\rangle(\lambda_{<n>}+(-1)^{\nu}\mu_{<n>})+ +(\langle n^{k+1-\nu}\rangle-\langle n^{k-\nu}\rangle\langle n\rangle)(\lambda'_{<n>}+(-1)^{\nu}\mu'_{<n>})\Big] \qquad (D.4.4)$$

Hence the equation for the k-th moment contains only moments up to the k-th moment and the hierarchy of (approximate) equations given by (D.4.4) can be solved step by step.

4.2 Bilinear One-Variable Birth-Death Processes

In bilinear processes the transition moments are given by

$$\lambda_n = \overset{0}{\lambda} + \overset{1}{\lambda} n + \overset{2}{\lambda} n^2$$

$$\mu_n = \overset{0}{\mu} + \overset{1}{\mu} n + \overset{2}{\mu} n^2 \qquad (D.4.5)$$

The equation (D.2.13) for the first moment in this case contains the second moment, variance respectively:

$$\frac{d\langle n\rangle}{dt} = (\overset{0}{\lambda} - \overset{0}{\mu}) + (\overset{1}{\lambda} - \overset{1}{\mu})\langle n\rangle + (\overset{2}{\lambda} - \overset{2}{\mu})\langle n^2\rangle$$

$$= (\overset{0}{\lambda} - \overset{0}{\mu}) + (\overset{1}{\lambda} - \overset{1}{\mu})\langle n\rangle + (\overset{2}{\lambda} - \overset{2}{\mu})(\langle n\rangle^2 + \sigma^2) \qquad (D.4.6)$$

Similarly the first terms on the right-hand side of equation (D.2.16) for σ^2 contain the third moment. Linearization according to (D.4.1) yields the approximate equations

$$\frac{d\langle n\rangle}{dt} = (\overset{0}{\lambda} - \overset{0}{\mu}) + (\overset{1}{\lambda} - \overset{1}{\mu})\langle n\rangle + (\overset{2}{\lambda} - \overset{2}{\mu})\langle n\rangle^2 \qquad (D.4.7)$$

$$\frac{d\sigma^2}{dt} = (\overset{0}{\lambda} + \overset{0}{\mu}) + (\overset{1}{\lambda} + \overset{1}{\mu})\langle n\rangle + (\overset{2}{\lambda} + \overset{2}{\mu})\langle n\rangle^2$$

$$+ 2(\overset{1}{\lambda} - \overset{1}{\mu})\sigma^2 + 4(\overset{2}{\lambda} - \overset{2}{\mu})\langle n\rangle\,\sigma^2 \qquad (D.4.8)$$

Comparison of (D.4.7) with (D.4.6) shows that the described approximation procedure based on a linearization (D.4.1) of the moments around $\langle n\rangle(t)$ means that σ^2 can be neglected compared with $\langle n\rangle^2$. An inspection of the higher moments yields corresponding results.

4.3 Bimolecular Reactions: The Gramicidin Channel

As an example for bilinear processes we shortly discuss the bimolecular reaction

$$A + B \underset{k_D}{\overset{k_R}{\rightleftharpoons}} D \qquad (D.4.9)$$

By this reaction can be described the kinetics of the polipeptide gramicidin A, which has been used as a model compound for the study of ion-transport through channels(Hladky a. Haydon, 1972, Bamberg a. Läuger, 1973). It has been shown that a single channel is formed by association of two gramicidin monomers to a dimer. Usually the channel forming reaction kinetics for gramicidin A are described by

$$A + A \underset{k_D}{\overset{k_R}{\rightleftharpoons}} D \qquad (D.4.10)$$

But we want to distinct between monomers A and B at different sides of the membranes, because there is evidence that a dimer is formed by monomers at different sides of the membranes. (D.4.10) may be treated analogously. Hence (under the assumption of fast transport) the measured current is the number of dimers times the single channel conductance. A paper concerning the experimental and theoretical analysis of current fluctuations generated by gramicidin channels at transient states is being prepared (Kolb, Junges a. Frehland, to be published).

Assuming constant numbers A, B of gramicidin molecules on the left, right sides of the membrane, the following conservation relations are valid:

$$N_A + N_D = C_A$$

$$N_B + N_D = C_B \qquad (D.4.11)$$

(N_A, N_B, N_D: numbers of monomers, dimers respectively). The usual (macroscopic) kinetic equations describing the time dependence of N_A, N_B, N_C are:

$$-\frac{dN_A}{dt} = -\frac{d\,N_B}{dt} = +\frac{d\,N_D}{dt} = +k_R N_A N_B - k_D N_D \qquad (D.4.12)$$

We are interested in a stochastic description of N_D by a birth-death process. The decisive step is the derivation of the coefficients $\overset{o}{\lambda},\overset{1}{\lambda},\overset{2}{\lambda},\overset{o}{\mu},\overset{1}{\mu},\overset{2}{\mu}$ in (D.4.5)-(D.4.8) . The first term in (D.4.12) comes from the birth process and the second one from the death process. The birth transition moment $k_R N_A N_B$ contains N_A and N_B, which can be eliminated with the use of the conservation relations (D.4.11). After some manipulations one gets

$$\frac{d\,N_D}{dt} = k_R(C_A-N_D)(C_B-N_D) - k_D N_D =$$

$$= k_R C_A C_B - k_R(C_A+C_B)\,N_D + k_R\,N_D^2 - k_D N_D \qquad (D.4.13)$$

In the approximation described above,(D.4.13) can be considered as the equation for the first moment. The transition moments are

$$\overset{o}{\lambda} = k_R C_A C_B \quad , \quad \overset{1}{\lambda} = -\,k_R(C_A+C_B), \quad \overset{2}{\lambda} = k_R$$

$$\overset{o}{\mu} = o \quad , \quad \overset{1}{\mu} = k_D \quad , \quad \overset{2}{\mu} = o \qquad (D.4.14)$$

With (D.4.14) and replacing $\langle n\rangle$ by $\langle N_c\rangle$ in (D.4.7) and (D.4.8) one gets the approximate equations for the first moment and the variance.

In case the kinetics are described by (D.4.10) the transitions can be derived analogously. The result corresponds to (D.4.14) in the case of equal concentrations $C_A=C_B$. The conservation relations now are

$$2\,N_A + N_D = N \ , \qquad (D.4.15)$$

where N corresponds to (C_A+C_B),

and the kinetic equation for N_D becomes

$$\frac{d\,N_D}{dt} = 4k_R N_D^2 - (4k_R N + k_D) N_D + k_R N^2 \qquad \text{(D.4.16)}$$

The transition moments are

$$\overset{o}{\lambda} = k_R N^2 \;,\; \overset{1}{\lambda} = -4k_R N \;,\; \overset{2}{\lambda} = 4k_R$$

$$\overset{o}{\mu} = o \;,\; \overset{1}{\mu} = k_D \;,\; \overset{2}{\mu} = o \qquad \text{(D.4.17)}$$

Comparison with (D.4.14) for $C_A=C_B=\frac{N}{2}$ shows that in this case both reaction mechanisms yield the same results, if one replaces $4k_R$ in (D.4.17) by k_R in (D.4.14).

5. Transport Noise at Transient States

In this last section we briefly show that the approach to transport fluctuations around steady states presented in section 4 of part C can be applied also for the treatment of transient states. In the derivation of (C.4.15) we have to replace t=o by t and t by t+S. (C.4.13) now becomes

$$\langle \phi_{\mu\nu}(t)\ \phi_{\kappa\rho}(t+S)\rangle = \lambda_{\mu\nu}(t)\quad \lambda_{\kappa\rho}(t+S)_{(\phi_{\mu\nu}(t)=h)} \tag{D.5.1}$$

with C.4.7. $\lambda_{\mu\nu}(t)$ is the expected flux $\langle\phi_{\mu\nu}(t)\rangle$ and in the limit $(h\to\infty,\ \tau\to o)$ $\lambda_{\kappa\rho}(t+S)$ in (D.5.1) is the expected flux on condition that at time t a transition $\nu\to\mu$ occurs, i.e. at t the system is in state μ. With the definition of the fundamental solutions to be the deviation from the steady state we get

$$\lim_{\substack{h\to\infty\\ \tau\to o}} \lambda_{\kappa\rho}(t+S)_{(\phi_{\mu\nu}(t)=h)} = M_{\kappa\rho}\Omega_{\rho\mu}(S) + \phi^{S}_{\kappa\rho} \tag{D.5.2}$$

Then the time correlations between individual fluxes are in extension of (C.4.15)

$$\langle\phi_{\mu\nu}(t)\phi_{\kappa\rho}(t+S)\rangle = \langle\phi_{\mu\nu}(t)\rangle\delta_{\mu\nu,\kappa\rho}\delta(S) + \langle\phi_{\mu\nu}(t)\rangle M_{\kappa\rho}\Omega_{\rho\mu}(S) + \langle\phi_{\mu\nu}(t)\rangle\ \phi^{S}_{\kappa\rho} \tag{D.5.3}$$

Then the autocorrelation function $C_J(t)$ of current fluctuations is

$$C_J(t,t+S) = \sum_{\mu\nu} \gamma^2_{\mu\nu}\langle\phi_{\mu\nu}(t)\rangle\cdot\delta(S) +$$

$$+ \sum_{\mu,\nu,\kappa,\rho} \gamma_{\mu\nu}\gamma_{\kappa\rho}\langle\phi_{\mu\nu}(t)\rangle \left[M_{\kappa\rho}\Omega_{\rho\mu}(S) + \phi^{S}_{\kappa\rho} \right] \qquad (D.5.4)$$

With the use of (D.5.4) all transport models discussed in part C can be analyzed at transient states. Because presently it is doubtful if a sensible experimental analysis can be performed we renounce to a further explicit discussion of special transport systems.

REFERENCES

Aschinger, G.A. (1975), Lecture notes in economics and mathematical systems, vol.113 (Springer, Berlin-Heidelberg-New York).

Baily, N.T.J. (1964), The Elements of Stochastic Processes with Applications to the Natural Sciences, (Wiley, New York).

Bak, T.A. (1959), Contributions to the theory of chemical kinetics (Fr.Baggs Kgl. Hofbogtrykkeri,Kobenhaven).

Bamberg E., Läuger P. (1973), Channel Formation Kinetics of Gramicidin A in Lipid Bilayer Membranes, J.Membrane Biol. 11, 177-194.

Bell, D.A. (1960), Electrical Noise (Van Nostrand, London).

Bendat, J.S. and Piersol, A.G. (1971), Random Data: Analysis and Measurement Procedures (Wiley, New York).

Bharucha-Reid, A.T. (1960), Elements of the Theory of Markow Processes and Their Applications.(McGraw-Hill, New York).

Bittel, H. and Storm L. (1971), Rauschen-Eine Einführung zum Verständnis elektrischer Schwankungserscheinungen (Springer, Berlin).

Braun, M. (1978), Differential Equations and Their Application (Springer, New York).

Callen, H.B., Welton T.A. (1951). Irreversibility and Generalized Noise, Phys.Rev.83, 34-40.

Callen, H.B. and Greene, R.F. (1952), On a Theorem of Irreversible Thermodynamics, Phys. Rev. 86, 702-710.

Casimir H.B.G. (1945), On Onsager's Principle of Microscopic Reversibility Rev. Modern Phys. 17, 343-350.

Chen. Y.D. (1975), Fluctuations and Noise in Kinetic Systems-Cycling Steady-State Models, J. Theor. Biol. 55, 229-243.

Chen. Y.D. (1975), Matrix method for fluctuations and noise in kinetic systems, Proc. Nat. Acad. Sci. 72, 3807-3811.

Chen, Y.D. (1978), Noise Analysis of Kinetic Systems and its Applications to Membrane Channels, Advances in Chemical Physics 37, 67-97.

Clay, J.R. (1978), Comparison of Ion Current Noise Predicted from Different Models of the Na-Channel Gating Mechanism in Nerve Membrane, J. Membrane Biol. 42, 215-227.

Conti F., Wanke E. (1975), Channel Noise in Membranes and Lipid Bilayers, Quart. Rev. Biophys. 8, 451-506.

Conti, F. Carbone E. (1981), Voltage Gated Channels and their Noise in Nerve and Muscle Membranes, in: Electrical Fields in Biological Membranes (Zimmermann, U., Benz, R., eds.; Springer, Berlin).

Doob, J.L. (1953), Stochastic Processes (Wiley, New York).

De Felice,L.J. (1977), Fluctuation Analysis in Neurobiology, Int. Rev. of Neurobiology 20, 169-208.

De Felice, L.J. (1981), Introduction to Membrane Noise (Plenum Press, New York).

Feller, W. (1950), An Introduction to Probability Theory and its Application I (Wiley, New York).

Fishman, H.M., Poussart, D.J.M., Moore, L.E. and Siebenga, E.(1977) K^+ Conduction and Admittance of Squid Axon, J. Membr. Biol. 32, 255-290.

Fishman, H.M., Poussart, D., Moore, L.E. (1979), Complex admittence of Na^+ conduction in squid axon, J. Membrane Biol., 50, 43-63.

Frehland E., Läuger P. (1974), Ion transport through pores: transient phenomena, J. theor. Biol. 47, 189-207.

Frehland E. (1978), Current noise Around Steady States in Discrete Transport Systems, Biophys. Chem. 8, 255-265; 10, 128.

Frehland E., Stephan W. (1979), Theory of Single-File Noise, Biochim. Biophys. Acta 553, 326-341.

Frehland E. (1979), Theory of Transport Noise in Membrane Channels with Open-Closed Kinetics, Biophys. Struct. Mech. 5, 91-106.

Frehland E., Faulhaber K.H. (1980), Nonequilibrium Ion Transport Through Pores, Biophys. Struct. Mech.7, 1-16.

Frehland E. (1980), Current Fluctuation in Discrete Transport Systems far from Equilibrium-Breakdown of the Fluctuation Dissipation Theorem, Biophys. Cem. 12, 63-71.

Frehland E. (1981), Transport Fluctuations Around Non Equilibrium Steady States In Discrete Systems. In: Proc. of the 6th. International Symposium on Noise in Physical Systems, Washington, 1981.

Frehland E. (1981), The Concept of Discrete Transport Systems - General Treatment and Discussion of Vectorial Transport Processes at Equilibrium and Nonequilibrium (submitted).

Frehland E., Junges R., H.-A.Kolb (1981), Variance of time and ensemble averaged current fluctuations generated by pore containing lipid membranes (to be published).

Goel N.S., Richter-Dyn N. (1974), Stochastic Models in Biology (Academic Press, New York).

de Groot S.R., Mazur P. (1962), Non-equilibrium Thermodynamics (North-Holland, Amsterdam).

Hearon J.Z. (1953), The Kinetics of Linear Systems with Special Reference to Periodic Reactions, Bulletin of Math. Biophys. 15, 121-141.

Heckmann K. (1965), Zur Theorie der "single-file"-Diffusion, I., Z. Phys. Chem. 44, 184-203.

Hladky S.B., Haydon D.A. (1972), Ion transfer across lipid membranes in the presence of gramicidin A, Biochim. Biophys. Acta 274, 294-312.

Heckmann,K.(1972), Single-File Diffusion, Biomembranes 3, 127-153.

Higgins J. (1967), Oscillating Reactions, Industrial and Engineering Chemistry, 59, 19-62,

Hille B., Schwarz W. (1978), Potassium channels as single-file pores, J. Gen. Physiol. 72, 409-442.

Hodgkin A.L., Keynes R.D. (1955), The Potassium permeability of giant nerve fibre, J.Physiol. 128, 61-88.

Jähnig F. and Richter P.H. (1976), Fluctuations in spatially homogeneous chemical steady states, J. Chem. Phys. 64, 4645-4656.

Jordan P.C. (1980), Current Noise in transport of hydrophobic ions through lipid bylayers including diffusion polarization in the aqueous phase, Biophys. Chem. 12, 1-11.

Junges R., Kolb H.-A. (1982), Noise analysis and relaxation experiments of transport of hydrophobic anions across lipid membranes at equilibrium and non-equilibrium. Biophys. Chem. (in press).

Kampen N.G. (1975), The Expansion of the Master Equation, Adv. in Chem. Phys., (Wiley, New York).

Karlin S. (1966), A first Course in Stochastic Processes (Academy Press, New York).

Keizer J. (1976), Fluctuations, stability and generalized state functions at nonequilibrium steady states, Chem. Phys. 65, 4431-4444.

Khintchine A. (1934), Korrelationstheorie der stationären stochastischen Prozesse, Math. Ann. 109, 604-615.

Kolb H.-A., Läuger P. (1977), Electrical noise from lipid bilayer membranes in the presence of hydrophobic ions, J. Memb. Biol. 37, 321-345.

Kolb H.-A., Läuger P. (1978), Spectral Analysis of Current Noise Generated by Carrier-Mediated Ion Transport, J. Membrane Biol. 41, 167-181.

Kolb H.-A., Frehland E. (1980), Noise-Current Generated by Carrier-Mediated Ion Transport at Non-Equilibrium, Biophys. Chem. 12, 21-34.

Kolb H.A. (1982), Electrical fluctuations in lipid bilayer membranes, Biophys. Struct. Mech. (in press).

Kubo R. (1965), Linear Response Theory of Irreversible Processes, in: Statistical Mechanics of Equilibrium and Non-Equilibrium (J. Meixner, ed., Ansterdam).

Kubo R. (1966), The fluctuation-dissipation theorem, Rep. Progr. Phys. (London) 29, 255-284.

Läuger P. (1973), Ion Transport through pores: a rate-theory analysis, Biochim. Biophys. Acta 311, 423-441.

Läuger P. (1975), Shot Noise in Ion Channels, Biochim. Biophys. Acta, 413, 1-10.

Läuger P. (1978), Transport Noise in Membranes - Current and Voltage Fluctuations at Equilibrium, Biochim. Biophys. Acta, 507, 337-349.

Läuger P., Stephan W., Frehland E. (1980), Fluctuations of Barrier Structure in Ionic Channels, Biochim. Biophys. Acta 602, 167-180.

Lax M. (1960), Fluctuations from the Nonequilibrium Steady State, Rev. Mod. Phys. 32, 25-64.

Mc Quarrie D.A. (1967), Stochastic Approach to Chemical Kinetics (Methuen, London).

Meixner J.,Reik H.G. (1959), Thermodynamik der irreversiblen Prozesse, in: Handbuch der Physik 3 (S.Flügge, ed.)

Moore L.E., Fishman H.M., Poussart D.J.M. (1980), Small-Signal Axons J. Membrane Biol. 54, 157-164.

Neher E. and Stevens C.F. (1977), Conductance Fluctuations and Ionic Pores in Membranes, Ann. Rev. Biophys. Bioeng. 6, 345-381.

Nyquist H. (1928), Thermal Agitation of Electric Charge in Conductors, Phys. Rev. 32, 110-113.

Pfeifer H. (1959), Elektronisches Rauschen 1, (Teubner, Leipzig).

Rickert H. (1964), Zur Diffusion durch eine lineare Kette (Single-File-Diffusion), Z. Phys. Chem. 46, 1-25.

Schnakenberg (1976), Network theory of microscopic and macroscopic behavior of master quation systems, Rev. Mod. Phys. 48, 571-585.

Schottky W. (1918), Über spontane Stromschwankungen in verschiedenen Elektrizitätsleitern, Ann. Physik 57, 541-567.

Sigworth F.J. (1977), Sodium channels in nerve apparently have two conductance states, Nature, 270, 265-266.

Sigworth F.J. (1981), Covariance of nonstationary sodium current fluctuations at the node of ranvier, Biophys. J. Biophys. Soc. 34, 111-133.

Stephan W. (1981), Oszillatorische Relaxationsphänomene als Strukturmerkmal zeitabhängiqigen Single-File-Transports(Dissertation, Konstanz).

Verveen, A.A., Deerksen H.E. (1965), Fluctuations in membrane potential of axons and the problem of coding, Kybernetik 2, 152-160.

Verveen, A.A. a. De Felice L.J. (1974), Membrane Noise, Prog. Biophys. Mol. Biol. 28, 189-265.

Vliet K.M. van, Fasset J.R. (1965), Fluctuations Due to Electronic Transitions and Transport in Solids, in: Fluctuations Phenomena in Solids (Burgess, ed., Academic Press, New York)

Wiener N. (1930), Generalized Harmonic Analysis, Acta.Math. 55, 117-258.

Ziel A. van der (1970), Noise: Sources, Characterization, Measurement, (Prentice-Hall, New Jersey).

Ziel A. van der (1976), Noise in Measurements (Wiley, London).

Zwolinsky B.J., Eyring H., Reese C.E. (1949), Diffusion and membrane permeability, J. Phys. Chem. 53, 1426-1453.

SUBJECT INDEX

Bio-mathematics

Managing Editor: S. A. Levin

Volume 8

A. T. Winfree

The Geometry of Biological Time

1979. 290 figures. XIV, 530 pages
ISBN 3-540-09373-7

The widespread appearance of periodic patterns in nature reveals that many living organisms are communities of biological clocks. This landmark text investigates, and explains in mathematical terms, periodic processes in living systems and in their non-living analogues. Its lively presentation (including many drawings), timely perspective and unique bibliography will make it rewarding reading for students and researchers in many disciplines.

Volume 9

W. J. Ewens

Mathematical Population Genetics

1979. 4 figures, 17 tables. XII, 325 pages
ISBN 3-540-09577-2

This graduate level monograph considers the mathematical theory of population genetics, emphasizing aspects relevant to evolutionary studies. It contains a definitive and comprehensive discussion of relevant areas with references to the essential literature. The sound presentation and excellent exposition make this book a standard for population geneticists interested in the mathematical foundations of their subject as well as for mathematicians involved with genetic evolutionary processes.

Volume 10

A. Okubo

Diffusion and Ecological Problems: Mathematical Models

1980. 114 figures, 6 tables. XIII, 254 pages
ISBN 3-540-09620-5

This is the first comprehensive book on mathematical models of diffusion in an ecological context. Directed towards applied mathematicians, physicists and biologists, it gives a sound, biologically oriented treatment of the mathematics and physics of diffusion.

Springer-Verlag
Berlin
Heidelberg
New York

Lecture Notes in Biomathematics

Vol. 26: M. B. Katz, Questions of Uniqueness and Resolution in Reconstruction from Projections. IX, 175 pages. 1978.

Vol. 27: N. MacDonald, Time Lags in Biological Models. VII, 112 pages. 1978.

Vol. 28: P. C. Fife, Mathematical Aspects of Reacting and Diffusing Systems. IV, 185 pages. 1979.

Vol. 29: Kinetic Logic – A Boolean Approach to the Analysis of Complex Regulatory Systems. Proceedings, 1977. Edited by R. Thomas. XIII, 507 pages. 1979.

Vol. 30: M. Eisen, Mathematical Models in Cell Biology and Cancer Chemotherapy. IX, 431 pages. 1979.

Vol. 31: E. Akin, The Geometry of Population Genetics. IV, 205 pages. 1979.

Vol. 32: Systems Theory in Immunology. Proceedings, 1978. Edited by G. Bruni et al. XI, 273 pages. 1979.

Vol. 33: Mathematical Modelling in Biology and Ecology. Proceedings, 1979. Edited by W. M. Getz. VIII, 355 pages. 1980.

Vol. 34: R. Collins, T. J. van der Werff, Mathematical Models of the Dynamics of the Human Eye. VII, 99 pages. 1980.

Vol. 35: U. an der Heiden, Analysis of Neural Networks. X, 159 pages. 1980.

Vol. 36: A. Wörz-Busekros, Algebras in Genetics. VI, 237 pages. 1980.

Vol. 37: T. Ohta, Evolution and Variation of Multigene Families. VIII, 131 pages. 1980.

Vol. 38: Biological Growth and Spread: Mathematical Theories and Applications. Proceedings, 1979. Edited by W. Jäger, H. Rost and P. Tautu. XI, 511 pages. 1980.

Vol. 39: Vito Volterra Symposium on Mathematical Models in Biology. Proceedings, 1979. Edited by C. Barigozzi. VI, 417 pages. 1980.

Vol. 40: Renewable Resource Management. Proceedings, 1980. Edited by T. Vincent and J. Skowronski. XII, 236 pages. 1981.

Vol. 41: Modèles Mathématiques en Biologie. Proceedings, 1978. Edited by C. Chevalet and A. Micali. XIV, 219 pages. 1981.

Vol. 42: G. W. Swan, Optimization of Human Cancer Radiotherapy. VIII, 282 pages. 1981.

Vol. 43: Mathematical Modeling of the Hearing Process. Proceedings, 1980. Edited by M. H. Holmes and L. A. Rubenfeld. V, 104 pages. 1981.

Vol. 44: Recognition of Pattern and Form. Proceedings, 1979. Edited by D. G. Albrecht. III, 226 pages. 1982.

Vol. 45: Competition and Cooperation in Neutral Nets. Proceedings, 1982. Edited by S. Amari and M.A. Arbib. XIV, 441 pages. 1982.

Vol. 46: E. Walter, Identifiability of State Space Models with applications to transformation systems. VIII, 202 pages. 1982.

Vol. 47: E. Frehland, Stochastic Transport Processes in Discrete Biological Systems. VIII, 169 pages. 1982.